SpringerBriefs in Energy

SpringerBriefs in Energy presents concise summaries of cutting-edge research and practical applications in all aspects of Energy. Featuring compact volumes of 50 to 125 pages, the series covers a range of content from professional to academic. Typical topics might include:

- A snapshot of a hot or emerging topic
- A contextual literature review
- A timely report of state-of-the art analytical techniques
- An in-depth case study
- A presentation of core concepts that students must understand in order to make independent contributions.

Briefs allow authors to present their ideas and readers to absorb them with minimal time investment.

Briefs will be published as part of Springer's eBook collection, with millions of users worldwide. In addition, Briefs will be available for individual print and electronic purchase. Briefs are characterized by fast, global electronic dissemination, standard publishing contracts, easy-to-use manuscript preparation and formatting guidelines, and expedited production schedules. We aim for publication 8–12 weeks after acceptance.

Both solicited and unsolicited manuscripts are considered for publication in this series. Briefs can also arise from the scale up of a planned chapter. Instead of simply contributing to an edited volume, the author gets an authored book with the space necessary to provide more data, fundamentals and background on the subject, methodology, future outlook, etc.

SpringerBriefs in Energy contains a distinct subseries focusing on Energy Analysis and edited by Charles Hall, State University of New York. Books for this subseries will emphasize quantitative accounting of energy use and availability, including the potential and limitations of new technologies in terms of energy returned on energy invested. The second distinct subseries connected to SpringerBriefs in Energy, entitled Computational Modeling of Energy Systems, is edited by Thomas Nagel, and Haibing Shao, Helmholtz Centre for Environmental Research - UFZ, Leipzig, Germany. This sub-series publishes titles focusing on the role that computer-aided engineering (CAE) plays in advancing various engineering sectors, particularly in the context of transforming energy systems towards renewable sources, decentralized landscapes, and smart grids.

All Springer brief titles should undergo standard single-blind peer-review to ensure high scientific quality by at least two experts in the field.

Isabel C. Gil-García · Adela Ramos-Escudero ·
Luis Serrano-Gómez · Ana Fernández-Guillamón

Smart Choices for Clean Energy

Multi-criteria Decision-Making
and Geographical Information Systems
in Renewable Energy Sources Planning

Isabel C. Gil-García
Faculty of Engineering, Distance University
of Madrid
Collado Villalba, Madrid, Spain

Luis Serrano-Gómez
Department of Electrical, Electronic,
Automatic and Communications
Engineering
University of Castilla-La Mancha
Albacete, Spain

Adela Ramos-Escudero
Department of Applied Physics
Polytechnic University of Cartagena
Cartagena, Murcia, Spain

Ana Fernández-Guillamón
Department of Applied Mechanics
and Projects Engineering
University of Castilla-La Mancha
Albacete, Spain

ISSN 2191-5520   ISSN 2191-5539 (electronic)
SpringerBriefs in Energy
ISBN 978-3-031-90205-5   ISBN 978-3-031-90206-2 (eBook)
https://doi.org/10.1007/978-3-031-90206-2

© The Author(s), under exclusive license to Springer Nature Switzerland AG 2025

This work is subject to copyright. All rights are solely and exclusively licensed by the Publisher, whether the whole or part of the material is concerned, specifically the rights of translation, reprinting, reuse of illustrations, recitation, broadcasting, reproduction on microfilms or in any other physical way, and transmission or information storage and retrieval, electronic adaptation, computer software, or by similar or dissimilar methodology now known or hereafter developed.
The use of general descriptive names, registered names, trademarks, service marks, etc. in this publication does not imply, even in the absence of a specific statement, that such names are exempt from the relevant protective laws and regulations and therefore free for general use.
The publisher, the authors and the editors are safe to assume that the advice and information in this book are believed to be true and accurate at the date of publication. Neither the publisher nor the authors or the editors give a warranty, expressed or implied, with respect to the material contained herein or for any errors or omissions that may have been made. The publisher remains neutral with regard to jurisdictional claims in published maps and institutional affiliations.

This Springer imprint is published by the registered company Springer Nature Switzerland AG
The registered company address is: Gewerbestrasse 11, 6330 Cham, Switzerland

If disposing of this product, please recycle the paper.

This book is the result of a journey made possible by the unwavering support, companionship, and inspiration of those who have been by our side every step of the way. It is the final product of the talent, ideas, and dedication of all those involved, making this book a true collaborative work filled with meaning.

Ana Fernández-Guillamón offers her heartfelt gratitude to her grandfather, whose constant example of lifelong learning continues to inspire, even in the face of adversity.

Isabel C. Gil-García extends her deepest appreciation to her dog, a loyal friend and companion, whose silent presence during countless hours of writing provided the support she needed.

Foreword

The Greek myth of Phaethon tells the story of a young man who discovers that he is the son of Helios, the god of the Sun. Eager to prove his divine lineage, Phaethon claims the privilege of driving the solar chariot. Helios warns him that no mortal is capable of controlling the celestial horses that pull the Sun's chariot. However, Phaethon persists until the Sun god reluctantly grants his request. After rising proudly into the sky, Phaethon ultimately loses control of the chariot. It veers off course, scorching all vegetation, causing countless deaths, and threatening to set the entire Earth ablaze. Zeus intervenes, striking Phaethon with a thunderbolt. The presumptuous mortal falls from the heavens such as a blazing star, consumed by fire. His blazing form plummeted into the river Eridanus, often identified with the current Po River in Italy. The gods reclaim control of the sky and succeed in saving the world. Throughout human history, numerous traditions have asserted that a fundamental and perilous flaw of humanity and our very nature is the tendency to seek dominion over powers and resources that we are incapable of controlling. However, beyond individual greed and arrogance, our true power arises from collaboration and cooperation among a great number of people. In my opinion, this is our true challenge and the mission entrusted to us—the collective and cooperative effort that enables advancements once thought unimaginable just a few decades ago. From this collaborative and cooperative perspective, the pursuit of solutions to achieve the optimal use of our available resources becomes meaningful—without provoking the wrath of the gods or relying on arrogance and greed, considered as inherently human as they have been intertwined with our existence on Earth. We must continue developing tools for the service of humanity, such as those described in this book, preventing that algorithms could take control in a dystopian future. Decisions must always be conducted towards the pursuit of the common good, and our power should be based on respect for the environment and respect for our own future. It is within this context where the ideas, solutions, and approaches discussed in this book will find their full meaning.

Cartagena, Colombia Angel Molina-Garcia
February 2025

Competing Interests The authors have no competing interests to declare that are relevant to the content of this manuscript.

Contents

1 Energy Sector 1
   1.1 Introduction 1
   1.2 Energy Sources 3
   1.3 Renewable Energy Sources for Electricity Generation 6
   1.4 Evolution of Power Systems 8
   1.5 Purpose of This Book 12
   References 13

2 Multi-criteria Decision-Making 17
   2.1 Introduction to MCDM 17
   2.2 Weighting the Criteria 18
      2.2.1 AHP 18
      2.2.2 Entropy 22
      2.2.3 Compromised AHP + Entropy 24
   2.3 Ranking of Alternatives 26
      2.3.1 Weighted Sum Method 26
      2.3.2 TOPSIS 26
      2.3.3 VIKOR 29
      2.3.4 SIMUS 31
   2.4 Fuzzy 33
      2.4.1 Fuzzy Sets 34
      2.4.2 Uncertainty: Contribution of Fuzzy Logic
            to Decision-Making 36
      2.4.3 Application Example: Software @Risk 36
      2.4.4 Advantages and Applications 43
   References 43

3 Geographical Information Systems 47
   3.1 GIS-Based Spatial Analysis for RES 47
   3.2 Geo-information for Building Institutional Capacity
       in Renewable Energy 50
   3.3 GIS-Based Decision Support 51

		3.3.1	Resource Potential Maps	52
		3.3.2	Technical Potential Maps	54
		3.3.3	Economic Potential Maps	54
		3.3.4	Market Potential Maps	54
		3.3.5	Conflict Use Maps	55
	3.4	GIS-Based Mapping Typical Workflow		56
	3.5	From Continental to Regional Scales: GIS-Based Studies of Renewable Energy Potential		57
		3.5.1	Continental Scale	58
		3.5.2	National to Regional Scale: MCDM-Supported Suitability Maps	58
		3.5.3	Local or Urban Scale	59
	References			59
4	**Integration of MCDM with GIS and Case Studies**			**63**
	4.1	Introduction		63
	4.2	Methodology		65
		4.2.1	Phase 1	65
		4.2.2	Phase 2	67
		4.2.3	Phase 3	67
	4.3	Application of GIS + MCDM		68
		4.3.1	Offshore Wind Energy	69
		4.3.2	Onshore Wind Energy	76
		4.3.3	Solar Photovoltaic	82
		4.3.4	Geothermal	87
	4.4	Conclusion		101
	References			101

Acronyms

AHP	Analytic Hierarchy Process
ANP	Analytic Network Process
ASHPs	Air-Source Heat Pumps
CAPEX	Capital Expenditure
CDD	Cooling Degree Days
CO_2	Carbon Dioxide
COP	Coefficient of Performance
CSV	Comma-Separated Values
DECEX	Decommissioning Expenditure
EU	European Union
ERM	Efficient Results Matrix
FAHP	Fuzzy Analytic Hierarchy Process
GHG	Greenhouse Gases
GHI	Global Horizontal Irradiance
GIS	Geographic Information Systems
GSHP	Ground Source Heat Pump
HDD	Heating Degree Days
H&C	Heating & Cooling
IS	International System
LCOE	Levelized Cost of Energy
MCDM	Multi-criteria Decision-Making
OPEX	Operational Expenditure
PCC	Point of Common Coupling
PDM	Project Dominance Matrix
PROMETHEE	Preference Ranking Organization Method for Enrichment of Evaluations
PV	Photovoltaic
RES	Renewable Energy Sources
RI	Risk Impact
RP	Risk Probability
SGE	Shallow Geothermal Energy

SIMUS	Sequential Interactive Modeling for Urban Systems
SMSPs	Spanish Maritime Spatial Planning
TOPSIS	Technique for Order of Preference by Similarity
VIKOR	VlseKriterijumska Optimizacija I Kompromisno Resenje
WAsP	Wind Atlas Analysis and Application Program
WTs	Wind Turbines

Chapter 1
Energy Sector

Abstract Electricity is essential in today's world, powering everything from homes and industries to communications and healthcare. It drives economic growth, improves quality of life, and supports innovation. However, traditional electricity production, largely reliant on fossil fuels, is a significant source of greenhouse gas emissions, which contribute to climate change. Rising global temperatures, extreme weather events, and environmental degradation are among the consequences of these emissions, highlighting the urgent need to transform how we produce and consume electricity. The shift to renewable energy sources, such as solar, wind, and hydropower, offers a solution to reduce carbon emissions and promote sustainable growth. Yet, this transition faces challenges, including technological limitations, financial costs, and the need for large-scale infrastructure updates. Additionally, renewable energy sources are often intermittent, requiring advanced storage and grid management systems. Despite these challenges, the move toward renewable energy is accelerating globally, driven by innovations in clean energy technology and increased public awareness of climate change impacts. As societies work to achieve energy sustainability, renewable energy emerges as a key element in creating a resilient and environmentally responsible energy future.

1.1 Introduction

Energy (E, measured in Jules [J], following the International System (IS)) is a physical magnitude associated to the ability to do any kind of work. There are many forms of energy, mainly gravitational, motion, chemical, heat, light, electrical, and chemical, being possible to convert from one form to another, as long as the principle of conservation of energy is fulfilled [22]. Power (P, measured in watts [W] according to the IS) is defined as the energy used (E) by a unit of time (t) [79]:

$$P = \frac{E}{t}. \tag{1.1}$$

© The Author(s), under exclusive license to Springer Nature Switzerland AG 2025
I. C. Gil García et al., *Smart Choices for Clean Energy*, SpringerBriefs in Energy, https://doi.org/10.1007/978-3-031-90206-2_1

From the power definition in Eq. (1.1), energy can be determined as:

$$E = P \cdot t. \tag{1.2}$$

Consequently, energy can also be measured in watt-hour [W h], commonly used in terms of electrical energy [60].

The Sun is the star that belongs to the Solar System. It is a yellow dwarf star, with a surface temperature of 5,780 K, composed essentially of hydrogen, helium, and a mixture of more than 100 chemical elements [40]. Atomic fusion reactions are constantly taking place in the Sun's core, transforming hydrogen atoms into helium atoms and releasing a large amount of energy [49]. Most of the energy received on the Earth comes from the Sun (in fact, more than 99.9%) [28]. It causes a series of phenomena on the atmosphere, water, and the Earth itself, such as the evaporation of surface water, wind, cloud formation, rain [7], etc. In addition, it is also responsible for all the processes that take place on the Earth; its heat and light are the basis for numerous chemical reactions essential for the development of plants and animals which, over the centuries, gave rise to fossil fuels (coal, oil, and natural gas). Along with the energy from the Sun, there is also the nuclear energy of some radioactive elements (uranium) present on the Earth; the gravitational energy of the Earth-Sun interaction, which can be harnessed from the movement it produces on water masses; and the hot magma of the Earth's interior, harnessed as geothermal energy [57]. Along with this, the Earth has a unique feature: a small part of its surface has life (in the form of plants and animals), which has been converting and storing the electromagnetic energy coming from the Sun through various mechanisms. This stored energy is known as biomass, and can be released by oxidation (combustion) at a rate equal to its storage. In addition, a small amount of this energy in the form of biomass has been stored for millions of years by being buried underground in incompletely oxidized conditions, giving rise to fossil fuels: coal, oil, natural gas, oil shale, tar sands [3].

The Sun's rays propagate through space in the form of electromagnetic waves. This phenomenon is called solar radiation, and it is responsible for the Earth receiving a total of $173 \cdot 10^{12}$ W in the exterior of the atmosphere, equivalent to 1, 367 W/m^2, which is called the solar constant [74]. The solar constant represents the amount of power received in the form of solar radiation per unit area, measured in the outer part of the Earth's atmosphere in a plane perpendicular to the Sun's rays. The value obtained from satellite measurements and currently accepted is 1, 367 W/m^2. However, as solar radiation passes through the atmosphere, it undergoes various processes that reduce the amount received by the Earth's surface due to its interaction with the various components of the atmosphere [30]. Thus, 30% is reflected into space; 50% is absorbed, heating the atmosphere and the Earth's surface, and being radiated back into space; the remaining 20% feeds the hydrological water cycle (19.76%), drives winds (0.18%) and fuels photosynthesis mechanisms (0.06%), as depicted in Fig. 1.1.

Fig. 1.1 Solar radiation

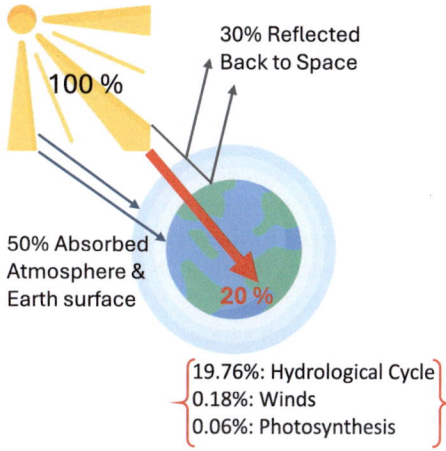

1.2 Energy Sources

An energy source is a natural system whose energy content can be converted into useful energy. It can be a physical body (such as fossil fuels) or be in diffuse form (such as solar, wind, or geothermal energy). The different energy sources are classified into two main groups: non-renewable and renewable [81].

- **Non-renewables**. These are those sources that are exhausted when their useful energy is used. Their quantity is limited, and they are not renewed in the short term (millions of years are necessary for their formation), so they will eventually run out. These sources include [68]:

  – Fossil fuels: coal, oil and natural gas.
  – Nuclear fuels: mainly uranium.

- **Renewables**. They are those sources whose potential is inexhaustible as they are constantly renewed. Therefore, they do not disappear when their useful energy is used. Renewable energy sources include, among others [26]:

  – Water (hydroelectric energy).
  – Wind (wind energy).
  – Solar radiation (solar energy).
  – Biomass.
  – Heat from the earth's interior (geothermal energy).
  – Seas and oceans (tidal and wave energy).

One of the most important problems of energy is related with how to convert the energy sources available into usable energy. Energy obtained directly from nature is primary energy. Primary energy, therefore, is energy that has not undergone any conversion process. This energy is generally not used directly, but is transformed into intermediate energy (mechanical, electrical, or thermal) to be transported to the consumer, who consumes it, for the most part, in the form of heat (thermal energy), work (mechanical energy), or electricity. The types of energy actually used by the consumer constitute the so-called end-use energies [76].

Primary energy consumption represents the total energy consumed in a country or region. It therefore indicates the amount of energy required to meet the consumption needs of the area under consideration. It includes energy consumed by the energy sector itself; transformation, transport, and distribution losses; and final energy consumed by users, which usually represents 60–70% of primary energy [5]. It also depends on various factors, such as the structure of the energy system (electrification, transport), the availability of resources for primary energy production (coal mines, oil fields, availability of resources for hydroelectric power plants, biofuel production, etc.) and the structure and development of the country's economy (less energy is consumed in economic recessions) [33]. Figure 1.2 shows the primary energy consumption according to the different countries of the world in 2023. As can be seen, this consumption is closely related to the level of development of the country, as well as to its land area. Thus, more developed areas such as Asia and the United States have higher consumption than Africa and South America, underdeveloped or developing continents.

The evolution of global primary energy consumption from 1,900 to 2,023, broken down by source, is illustrated in Figs. 1.3 (absolute data) and 1.4 (percentage).

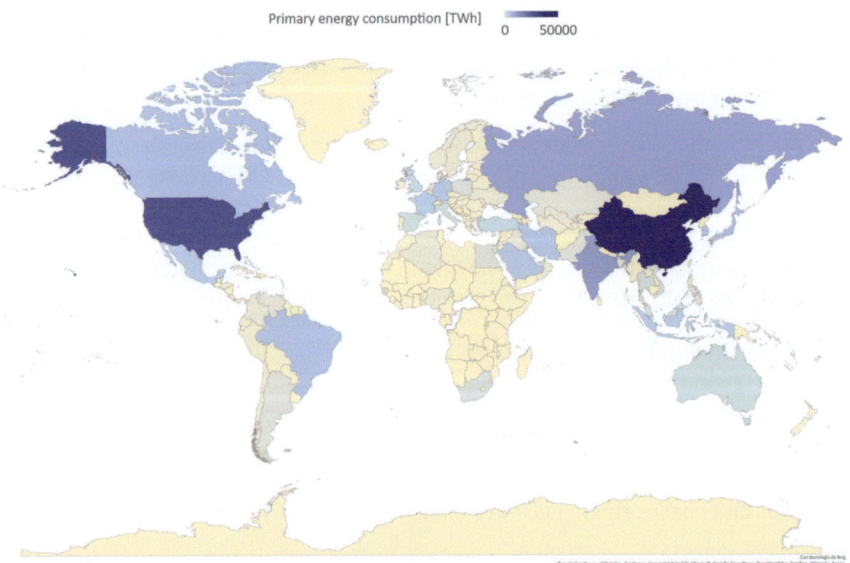

Fig. 1.2 Primary energy consumption by country in 2023. Own elaboration with data from Ritchie et al. [65]

1.2 Energy Sources

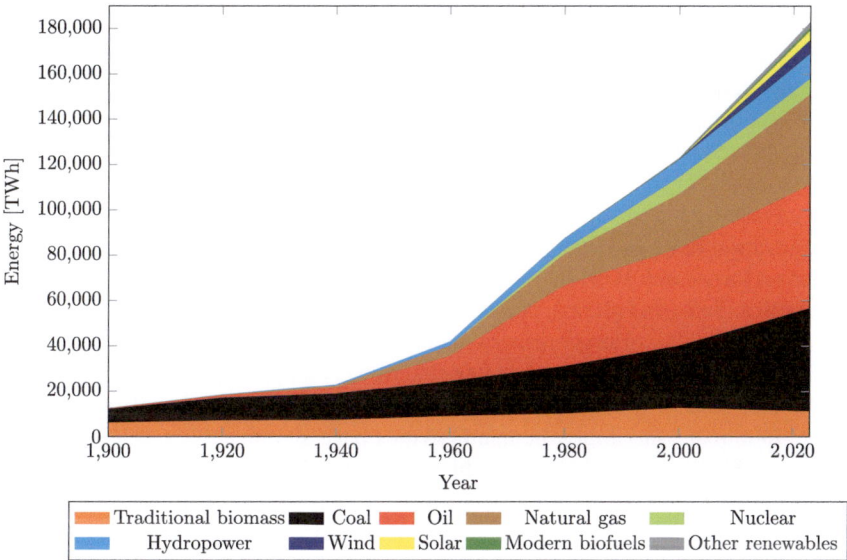

Fig. 1.3 Evolution of global primary energy consumption by source. Own elaboration with data from Ritchie et al. [65]

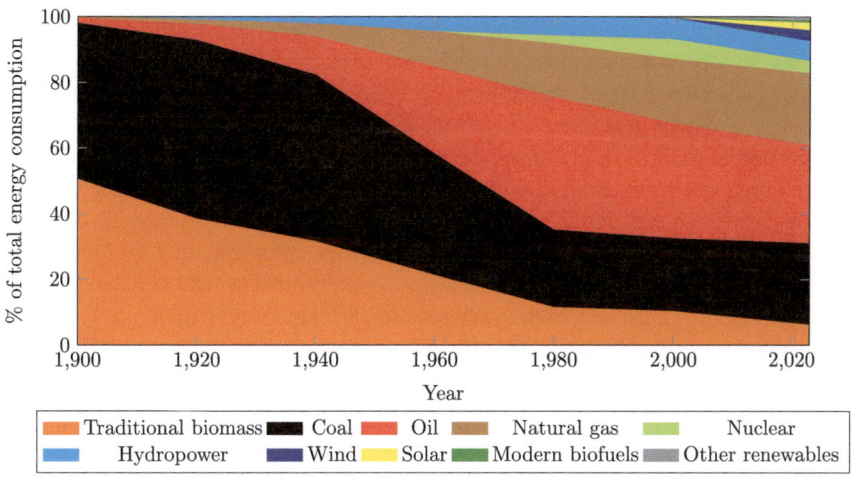

Fig. 1.4 Evolution of the percentage of global primary energy consumption by source. Own elaboration with data from Ritchie et al. [65]

Over the past century, energy consumption has grown more than fifteenfold, rising from 12,131 TW h in 1,900 to 183,230 TW h in 2,023. However, the vast majority of energy consumed today still comes from non-renewable sources, which account for 80% of total consumption. The Industrial Revolution and subsequent industrialization spurred technological progress that is now considered indispensable [52]; as a consequence, energy demand has suffered an exponential growth due to the technical advances in civilization and the increase of global population [55]. However, these advances have also resulted in significant environmental impacts, manifesting in numerous forms such as oil shortages, ocean pollution, water scarcity, species extinction (both animals and plants), deforestation and, most notably, climate change [23], one of the greatest worldwide environmental challenges.

Climate change, defined as the alteration of weather patterns, is primarily driven by the emission of greenhouse gases (GHG) [77]. Although natural factors contribute to climate variability, human activities have significantly accelerated these changes [75]. These, combined with historical events, have played a crucial role in shaping global energy priorities. The first oil crisis of the 1970s underscored the vulnerabilities of fossil fuel dependence, while the anti-nuclear movement—spurred by the Chernobyl disaster in the 1980s—further highlighted the need for alternative energy strategies. These factors collectively drove a growing interest in renewable energy sources (RES), which are seen as the most effective solution for achieving low-carbon development [6, 42].

The energy sector, encompassing electricity, transport, and heating and cooling, is a central focus of efforts to reduce GHG emissions [59]. Although electricity represents only about 20% of global primary energy consumption, nearly two-thirds of electricity is still generated from fossil fuels [1]. But transitioning to non-fossil energy sources, such as RES and nuclear power, offers a clear pathway to significantly reduce emissions from electricity generation [8]. Consequently, and in line with the introduction of international policies aimed at mitigating environmental degradation (such as the Kyoto Protocol (1997), the Doha Amendment (2012), and the Paris Agreement (2015) [46]), in recent years, governments worldwide have implemented numerous measures to promote sustainable electricity generation. Examples of these measures can be seen for different areas worldwide, such as the European Union [34], Algeria [72], Turkey [10], and some Asian countries (Japan, South Korea, and Taiwan) [18].

1.3 Renewable Energy Sources for Electricity Generation

The main RES technologies for electricity generation include:

- **Hydropower**: Hydropower is generated from any moving body of water, such as the flow of a river or water channelled through a conduit due to a height difference between two reservoirs [80]. Hydroelectric power plants harness this energy, which is determined by two key factors: flow rate and head. Water's hydraulic

1.3 Renewable Energy Sources for Electricity Generation

energy is converted into mechanical energy (in the form of torque and rotational speed) through a hydraulic turbine. The turbine's shaft is directly connected to an electric generator, where the mechanical energy is transformed into electricity [44]. Hydroelectric plants come in various sizes and designs, ranging from small installations on local rivers to large-scale dams with vast reservoirs [48].

- **Wind energy**: Wind energy is derived from the kinetic energy of moving air masses. This movement originates from temperature differences in the atmosphere, which are caused by varying intensities of solar radiation at both global and local scales. These temperature variations create upward and downward air currents, forming circulation patterns [24]. Wind power plants capture this energy by using wind turbines, which convert the mechanical rotational energy of the turbine into electricity through an electrical generator [17]. Unlike conventional power plants, where synchronous generators are typically used, wind power plants often employ asynchronous generators. These generators are more cost-effective and better suited to the fluctuating nature of wind conditions [71].
- **Solar energy**: Solar energy originates from the Sun in the form of radiation, which is a by-product of the Sun's internal nuclear reactions. These reactions, through fusion processes, convert nuclear energy into solar radiation [41]. Solar energy can be harnessed through both solar thermal systems and photovoltaic (PV) technology, with PV being the most advanced for electricity generation [4]:

  - PV power plants are made up of panels containing photovoltaic cells that convert solar radiation into direct current via the photovoltaic effect [70]. However, when exposed to sunlight, these panels heat up, leading to a reduction in efficiency. For every degree Celsius increase in temperature, panel efficiency drops slightly due to a decrease in open-circuit voltage, resulting in an overall efficiency loss of approximately 0.4–0.5% per degree [54].
  - Solar thermal power plants use concentrating collectors to focus sunlight from a large area onto a small, blackened receiver, significantly increasing the intensity of light to produce high temperatures [67]. Arrays of mirrors or lenses can concentrate enough sunlight to heat the target to temperatures exceeding 2000 °C. This heat is then used to run a boiler, generating steam to power a steam turbine [56].

  A key distinction between the two technologies lies in their electricity generation capabilities. Solar thermal plants, similar to conventional thermal power plants, have the ability to regulate electricity output, provided there is sufficient sunlight [43]. Additionally, many solar thermal systems incorporate thermal energy storage, which allows them to store excess heat and make the intermittent solar resource more controllable [61], unlike PV systems that lack such storage capabilities unless batteries or other additional energy storage systems are installed [13].

- **Bioenergy**: Bioenergy is a renewable energy source derived from organic materials, including agricultural and forestry residues, organic waste, and dedicated energy crops [38]. It is considered renewable for two main reasons: it comes from

sources that naturally regenerate over short cycles, and the plant-based fuels used contribute to the carbon cycle, helping to mitigate the impact of emissions [39].
- **Geothermal energy**: Geothermal energy is a renewable resource that harnesses heat from within the Earth to generate electricity or provide heating [19]. For electricity generation, geothermal power plants use this heat to warm water or another working fluid, driving turbines in either an open or closed cycle [53].
- **Ocean energy**: Ocean energy is an inexhaustible resource and is emerging as a promising source for supplying energy needs in a sustainable and reliable manner [50]. Within this context, two main technologies stand out: wave energy and tidal energy:

  - Wave energy is derived from the movement of waves on the surface of the sea. Waves carry enormous amounts of kinetic energy, which can be harnessed and captured for conversion into electrical energy. This is achieved by means of floating devices that move with the swaying of the waves, which generates mechanical energy that is subsequently transformed into electricity [2].
  - Tidal energy takes advantage of the rise and fall of the tides caused by the gravitational forces of the moon. Tidal installations employ weirs or turbines that harness the ebb and flow of water during these movements to convert, as in the case of wave power, mechanical energy into electricity [20].

1.4 Evolution of Power Systems

Apart from the inexhaustible potential of RES [35], they can address two critical challenges of power systems. The first one is related to the reduction of GHG emissions, particularly CO_2, as the combustion of fossil fuels is the primary driver of such emissions [45, 78]. In fact, according to some authors [21, 66], the electricity sector must be decarbonized by 2050 to meet the Paris Agreement's goal of limiting global temperature rise to below 2 °C above pre-industrial levels [11]. The second challenge is energy security, specifically reducing reliance on fuel imports from other countries [58]. The oil crises of the 1970s made countries acutely aware of the risks posed by supply disruptions, highlighting the importance of energy independence [29]. Dependence on foreign fuel supplies ties electricity security to factors such as market liberalization, political stability, and international relations, making energy supply vulnerable to external shocks [15]. Thus, minimizing fuel dependence enhances the stability of electricity supply [69].

As a consequence, in recent decades, power systems have evolved by gradually replacing polluting generation units (those fuelled by fossil fuels and nuclear energy) with RES [73], as shown in Figs. 1.5 (absolute data) and 1.6 (percentage). In 1,985, the global fossil fuels and nuclear power plants electrical generation accounted for nearly 80%, and have reduced around 10% in the last 40 years. This difference is related to the increase of RES integration in electricity generation, mainly solar and wind power, and others RES (such as geothermal and bioenergy) in small proportions.

1.4 Evolution of Power Systems

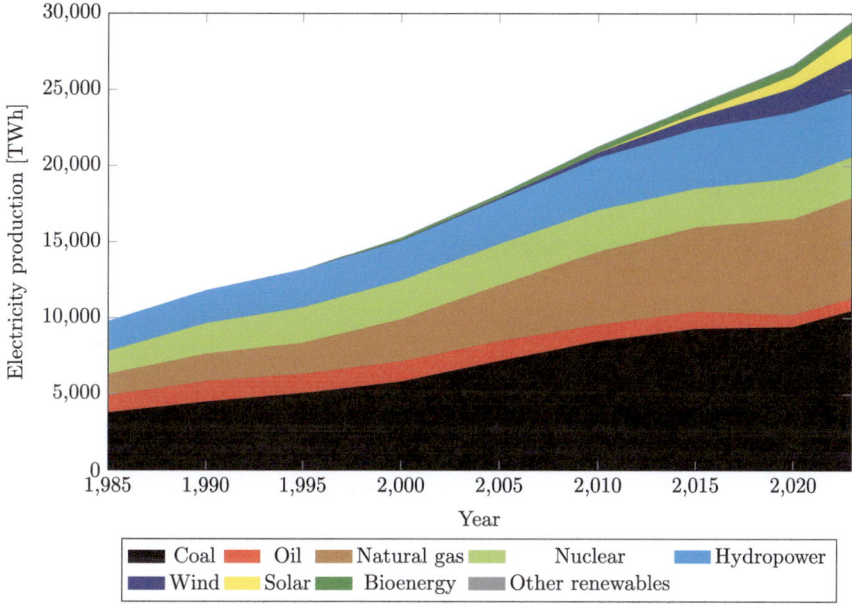

Fig. 1.5 Evolution of global electricity production by source. Own elaboration with data from Ritchie and Rosado [64]

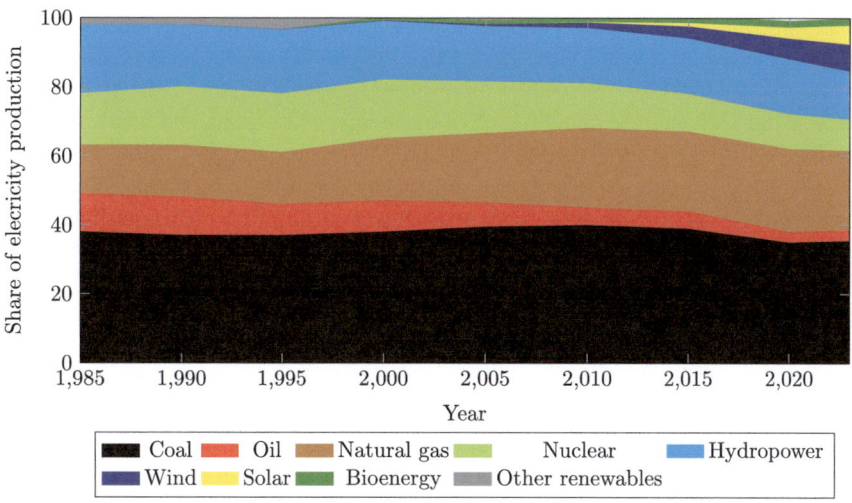

Fig. 1.6 Evolution of share of electricity production by source. Own elaboration with data from Ritchie and Rosado [64]

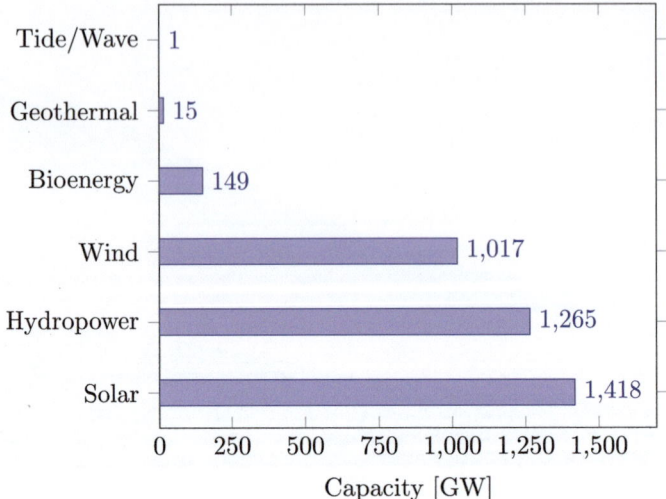

Fig. 1.7 RES capacity worldwide in 2023, by source. Own elaboration with data from International Renewable Energy Agency [36]

The installed capacity in 2,023 of the different RES presented in Sect. 1.3 is depicted in Fig. 1.7. The installed capacity shows solar energy as the most abundant, followed by hydropower and wind, while geothermal, bioenergy, and tide/wave sources are less developed. Considering that the global capacity including non-RES sources was 7,916 GW, renewables represented nearly 50% (solar 18%, hydropower 16%, wind 13%, and the combination of bioenergy, geothermal, and tide/wave 2%).

Figure 1.8 shows the electricity generation by source in 2,022. Similarly to the capacity, it highlights the dominance of the same three sources (solar, hydropower, and wind). However, in this case, hydropower is the RES that contributed the most, followed by wind and solar energy. The global electricity generation was 28,843.50 TW h. In this way, RES provided approximately 30% with 15% coming from hydropower, 7% from wind, 4.5% from solar, and 2.5% from the rest RES considered.

The comparison between installed capacity and electricity generation reveals important differences in the performance and utilization of RES. While RES account for nearly 50% of global installed capacity, they contribute only about 30% of global electricity generation. This discrepancy highlights variations in capacity factors (CF) among energy sources; this factor is defined as the ratio between the electricity produced by a power plant in a given period t—normally, a year—($E_{gen,t}$) and the electricity that could have produced if it had operated during the same period t at its nominal capacity P_{nom}, see Eq. (1.3).

$$CF = \frac{E_{gen,t}}{P_{nom} \cdot t} \tag{1.3}$$

1.4 Evolution of Power Systems

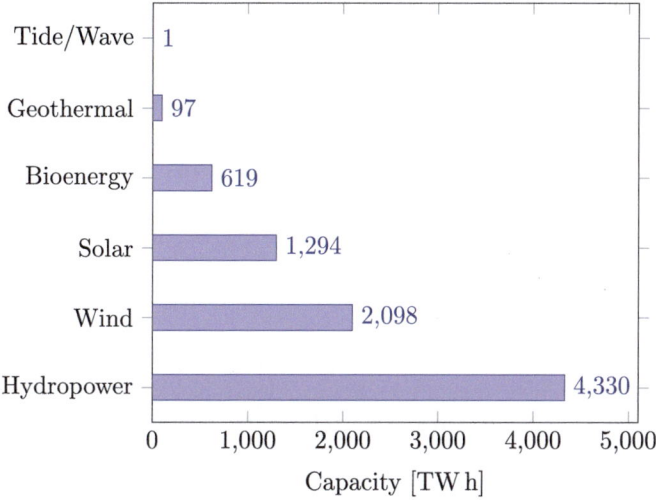

Fig. 1.8 RES electricity generation worldwide in 2022, by source. Own elaboration with data from International Renewable Energy Agency [36]

Table 1.1 Capacity factor values of the different power plants. Own elaboration from Bolson et al. [12]

	Biomass	Fossil	GEO[a]	HYD[b]	Nuclear	PV	Wind
CF [–]	0.53	0.48	0.76	0.46	0.84	0.12	0.23

[a]Geothermal
[b]Hydropower

Bolson et al. [12] analysed the CF of the different sources (both RES and non-RES) to analyse what is required to replace the current fossil fuels and nuclear plants by RES installations. The average CF of the different power plants are summarized in Table 1.1. As a consequence, replacing 1 W of fossil capacity requires approximately the installation of 1 W of biomass, 0.6 W of geothermal, 1 W of hydropower, 4 W of PV or 2 W of wind power. For nuclear power plants, these values would be nearly doubled.

A lower capacity factor correlates with lower generation [32]. This can be one of the main reasons why solar energy, for instance, leads among RES in installed capacity, but ranks third in generation, reflecting its lower utilization efficiency compared to hydropower and wind. Conversely, hydropower, with a slightly smaller share of capacity, contributes the highest percentage of RES generation due to its reliability and consistent output. These differences underscore the need to consider the capacity factor when planning energy strategies, ensuring a realistic understanding of the contributions and efficiencies of each source [9]. Moreover, and although the current share of RES in electricity generation remains relatively low, ambitious roadmaps have been proposed to significantly increase RES contributions. Examples of such roadmaps, projecting up to 2050, can be found in [25, 37, 62].

When planning the electricity transition, and together with the capacity factor, it is essential to consider the specific characteristics of each type of power plant, as these determine their potential roles within the power system [47]:

- **Base load plants**: These plants supply the portion of demand that is always present in the system. They are typically nuclear power plants and large thermal power plants.
- **Peak load plants**: These plants operate to meet spikes in demand. They are characterized by high flexibility and manoeuvrability. Typically, they include hydroelectric power plants and combined-cycle power plants, operating for only a few hours each day.
- **Regulation plants**: These plants are used to respond to fluctuations in demand. Like peak load plants, they are highly manoeuvrable. Examples include hydroelectric power plants, coal-fired power plants, and combined-cycle plants.
- **Reserve plants**: These plants have high generation costs and only operate to replace others that are offline due to maintenance or breakdowns. They are usually oil-fired or coal-fired thermal power plants.
- **Emergency plants**: These are small, mobile units deployed to areas without supply due to power outages. They are typically diesel generator sets.
- **Pumped storage plants**: These are hydroelectric plants located between two reservoirs. They store energy during periods of low demand by pumping water from a lower reservoir to an upper one, effectively storing energy as potential energy. This process consumes electricity during off-peak hours when demand is low. Pumped storage plants help smooth the system's load curve. When demand is high (peak hours), these plants generate electricity in the same way as other hydroelectric power plants.

However, both wind and PV, which are the most developed RES and used RES (but hydropower, refer to Figs. 1.7 and 1.8), are difficult to be considered in any of these categories. In fact, wind and PV energy generation share a critical limitation: their output directly depends on instantaneous primary energy availability (mainly wind speed and solar irradiation) [27]. As a consequence, it is essential to carefully select the optimal locations for RES power plants, ensuring that they are sited in areas with the best natural conditions for consistent and efficient generation Chatzoglou et al. [16]. This helps mitigate the variability of RES power plants and enhances their integration into the power grid.

1.5 Purpose of This Book

The integration of Geographic Information Systems (GIS) and Multi-Criteria Decision-Making (MCDM) methods is essential when planning new RES power plants, as these tools address the inherent variability and site-specific nature of RES [51]. GIS allows for the detailed analysis of spatial and environmental factors, such as solar irradiation, wind speed, proximity to transmission lines, and land-use constraints, ensuring that RES installations are located in areas with optimal natural and infrastructural conditions Resch et al. [63]. This not only enhances the capacity factors of these plants but also maximizes their generation potential, contributing to a more efficient and reliable energy system.

MCDM methods complement GIS by incorporating a range of technical, economic, social, and environmental criteria into the decision-making process [31]. By evaluating trade-offs between factors like energy yield, investment cost, environmental impact, and social acceptance, MCDM supports the selection of sites that balance competing priorities and align with long-term sustainability goals [14]. Together, GIS and MCDM enable a holistic approach to RES planning, ensuring that the transition to renewable energy systems is both effective and equitable, while minimizing risks associated with variability and inefficiencies in energy generation.

References

1. (2023) Electricity production from fossil fuels, nuclear and renewables, world. https://ourworldindata.org/grapher/elec-fossil-nuclear-renewables?
2. Aderinto T, Li H (2019) Review on power performance and efficiency of wave energy converters. Energies 12(22). https://doi.org/10.3390/en12224329
3. Ahluwalia V (2019) Energy and environment. The Energy and Resources Institute (TERI)
4. Ahmad L, Khordehgah N, Malinauskaite J, Jouhara H (2020) Recent advances and applications of solar photovoltaics and thermal technologies. Energy 207:118254. https://doi.org/10.1016/j.energy.2020.118254
5. Ahmad T, Zhang D (2020) A critical review of comparative global historical energy consumption and future demand: the story told so far. Energy Rep 6:1973–1991. https://doi.org/10.1016/j.egyr.2020.07.020
6. Aklin M, Urpelainen J (2018) Renewables: the politics of a global energy transition. MIT Press
7. Allen PA (2009) Earth surface processes. Wiley
8. Ang BW, Su B (2016) Carbon emission intensity in electricity production: a global analysis. Energy Policy 94:56–63
9. Babatunde OM, Munda JL, Hamam Y (2019) A comprehensive state-of-the-art survey on power generation expansion planning with intermittent renewable energy source and energy storage. Int J Energy Res 43(12):6078–6107
10. Baris K, Kucukali S (2012) Availability of renewable energy sources in turkey: current situation, potential, government policies and the EU perspective. Energy Policy 42:377–391
11. Bataille C, Åhman M, Neuhoff K, Nilsson LJ, Fischedick M, Lechtenböhmer S, Solano-Rodriquez B, Denis-Ryan A, Stiebert S, Waisman H et al (2018) A review of technology and policy deep decarbonization pathway options for making energy-intensive industry production consistent with the paris agreement. J Clean Prod 187:960–973
12. Bolson N, Prieto P, Patzek T (2022) Capacity factors for electrical power generation from renewable and nonrenewable sources. Proc Natl Acad Sci 119(52):e2205429119
13. Bullich-Massagué E, Cifuentes-García FJ, Glenny-Crende I, Cheah-Mañé M, Aragüés-Peñalba M, Díaz-González F, Gomis-Bellmunt O (2020) A review of energy storage technologies for large scale photovoltaic power plants. Appl Energy 274:115213
14. Calabrese A, Costa R, Levialdi N, Menichini T (2019) Integrating sustainability into strategic decision-making: a fuzzy ahp method for the selection of relevant sustainability issues. Technol Forecast Soc Chang 139:155–168

15. Chalvatzis KJ, Ioannidis A (2017) Energy supply security in the EU: benchmarking diversity and dependence of primary energy. Appl Energy 207:465–476
16. Chatzoglou P, Chatzoudes D, Petrakopoulou Z, Polychrou E (2018) Plant location factors: a field research. Opsearch 55:749–786
17. Chaudhuri A, Datta R, Kumar MP, Davim JP, Pramanik S (2022) Energy conversion strategies for wind energy system: electrical, mechanical and material aspects. Materials 15(3). https://doi.org/10.3390/ma15031232
18. Chen WM, Kim H, Yamaguchi H (2014) Renewable energy in eastern asia: renewable energy policy review and comparative swot analysis for promoting renewable energy in japan, south korea, and taiwan. Energy Policy 74:319–329
19. Chomać-Pierzecka E, Sobczak A, Soboń D (2022) The potential and development of the geothermal energy market in poland and the baltic states—selected aspects. Energies 15(11). https://doi.org/10.3390/en15114142
20. Chowdhury MS, Rahman KS, Selvanathan V, Nuthammachot N, Suklueng M, Mostafaeipour A, Habib A, Akhtaruzzaman M, Amin N, Techato K (2021) Current trends and prospects of tidal energy technology. Environ Dev Sustain 23(6):8179–8194. https://doi.org/10.1007/s10668-020-01013-4
21. Creutzig F, Agoston P, Goldschmidt JC, Luderer G, Nemet G, Pietzcker RC (2017) The underestimated potential of solar energy to mitigate climate change. Nat Energy 2(9):17140
22. Demirel Y (2012) Energy: production, conversion, storage, conservation, and coupling. Springer
23. Díaz Cordero G (2012) El cambio climático. Ciencia y sociedad
24. Emeis S (2018) Wind energy meteorology: atmospheric physics for wind power generation. Springer
25. ENTSO-E (2020) TYNDP 2020 – Scenario Report
26. Fernández-Guillamón A, Das K, Cutululis NA, Molina-García Á (2019) Offshore wind power integration into future power systems: overview and trends. J Mar Sci Eng 7(11):399
27. Fernández-Guillamón A, Gómez-Lázaro E, Molina-García Á (2020) Extensive frequency response and inertia analysis under high renewable energy source integration scenarios: application to the european interconnected power system. IET Renew Power Gener 14(15):2885–2896
28. Franjić S (2023) Unleashing sustainable energy: the sun, earth's largest and most powerfu source. J Sustain Dev 4(2):1–9
29. Gasser PE (2019) Quantifying electricity supply resilience of countries with multi-criteria decision analysis. PhD thesis, ETH Zurich
30. González JAC, Pérez RC, Santos AC, Gil MAC, Fernández EC (2009) Centrales de energías renovables: generación eléctrica con energías renovables. Pearson Educación Madrid
31. Greene R, Devillers R, Luther JE, Eddy BG (2011) Gis-based multiple-criteria decision analysis. Geogr Compass 5(6):412–432
32. Grubert E (2020) Same-plant trends in capacity factor and heat rate for us power plants, 2001–2018. IOP SciNotes 1(2):024007
33. Guevara Z, Henriques S, Sousa T (2021) Driving factors of differences in primary energy intensities of 14 european countries. Energy Policy 149:112090
34. Haas R, Panzer C, Resch G, Ragwitz M, Reece G, Held A (2011) A historical review of promotion strategies for electricity from renewable energy sources in EU countries. Renew Sustain Energy Rev 15(2):1003–1034
35. Halkos GE, Gkampoura EC (2020) Reviewing usage, potentials, and limitations of renewable energy sources. Energies 13(11). https://doi.org/10.3390/en13112906
36. International Renewable Energy Agency (2024) Electricity generation by energy source. https://www.irena.org/-/media/Files/IRENA/Agency/Publication/2024/Jul/Renewable_energy_highlights_FINAL_July_2024.pdf
37. IRENA E (2018) Renewable energy prospects for the european union. International Renewable Energy Agency (IRENA), European Commission (EC), Abu Dhabi
38. Janiszewska D, Ossowska L (2022) The role of agricultural biomass as a renewable energy source in european union countries. Energies 15(18). https://doi.org/10.3390/en15186756

References

39. Kalak T (2023) Potential use of industrial biomass waste as a sustainable energy source in the future. Energies 16(4). https://doi.org/10.3390/en16041783
40. Kaler JB (2016) From the sun to the stars. World Scientific
41. Kalogirou SA (2023) Solar energy engineering: processes and systems. Elsevier
42. Kamran M, Fazal MR (2021) Renewable energy conversion systems. Academic Press
43. Khan J, Arsalan MH (2016) Solar power technologies for sustainable electricity generation-a review. Renew Sustain Energy Rev 55:414–425
44. Ånund Killingtveit (2019) 8 - hydropower. In: Letcher TM (ed) Managing global warming. Academic Press, pp 265–315. https://doi.org/10.1016/B978-0-12-814104-5.00008-9
45. Lima M, Mendes L, Mothé G, Linhares F, de Castro M, Da Silva M, Sthel M (2020) Renewable energy in reducing greenhouse gas emissions: reaching the goals of the paris agreement in brazil. Environ Dev 33:100504
46. Lupiola AG (2016) El papel de la unión europea en la consecución de un acuerdo sustitutivo del protocolo de kioto: de bali a parís. Revista de estudios europeos 68:33–49
47. Machowski J, Lubosny Z, Bialek JW, Bumby JR (2020) Power system dynamics: stability and control. Wiley
48. Mayor B, Rodríguez-Muñoz I, Villarroya F, Montero E, López-Gunn E (2017) The role of large and small scale hydropower for energy and water security in the spanish duero basin. Sustainability 9(10). https://doi.org/10.3390/su9101807
49. McCracken G, Stott P (2013) Fusion: the energy of the universe. Academic Press
50. Melikoglu M (2018) Current status and future of ocean energy sources: a global review. Ocean Eng 148:563–573. https://doi.org/10.1016/j.oceaneng.2017.11.045
51. Mokarram M, Pham TM, Khooban MH (2022) A hybrid gis-mcdm approach for multi-level risk assessment and corresponding effective criteria in optimal solar power plant. Environ Sci Pollut Res 29(56):84661–84674
52. Moscoso ROA (2018) La industria y sus efectos en el cambio climático global. RECIAMUC 2(2):595–611
53. Moya D, Aldás C, Kaparaju P (2018) Geothermal energy: power plant technology and direct heat applications. Renew Sustain Energy Rev 94:889–901. https://doi.org/10.1016/j.rser.2018.06.047
54. Nwaigwe K, Mutabilwa P, Dintwa E (2019) An overview of solar power (pv systems) integration into electricity grids. Mater Sci Energy Technol 2(3):629–633. https://doi.org/10.1016/j.mset.2019.07.002
55. Olabi A, Abdelkareem MA (2022) Renewable energy and climate change. Renew Sustain Energy Rev 158:112111
56. Olabi AG, Shehata N, Maghrabie HM, Heikal LA, Abdelkareem MA, Rahman SMA, Shah SK, Sayed ET (2022) Progress in solar thermal systems and their role in achieving the sustainable development goals. Energies 15(24). https://doi.org/10.3390/en15249501
57. Orecchini F, Naso V (2011) Energy systems in the era of energy vectors: a key to define, analyze and design energy systems beyond fossil fuels. Springer
58. Pacesila M, Burcea SG, Colesca SE (2016) Analysis of renewable energies in european union. Renew Sustain Energy Rev 56:156–170
59. Papadis E, Tsatsaronis G (2020) Challenges in the decarbonization of the energy sector. Energy 205:118025
60. Patrick DR, Fardo SW, Fardo BW (2022) Electrical power systems technology. River Publishers
61. Pelay U, Luo L, Fan Y, Stitou D, Rood M (2017) Thermal energy storage systems for concentrated solar power plants. Renew Sustain Energy Rev 79:82–100
62. Ram M, Bogdanov D, Aghahosseini A, Oyewo S, Gulagi A, Child M, Fell HJ, Breyer C (2017) Global energy system based on 100% renewable energy—power sector. Lappeenranta University of Technology and Energy Watch Group, Lappeenranta, Finland
63. Resch B, Sagl G, Törnros T, Bachmaier A, Eggers JB, Herkel S, Narmsara S, Gündra H (2014) Gis-based planning and modeling for renewable energy: challenges and future research avenues. ISPRS Int J Geo-Inf 3(2):662–692

64. Ritchie H, Rosado P (2020) Electricity mix. Our World in Data https://ourworldindata.org/electricity-mix
65. Ritchie H, Rosado P, Roser M (2020) Energy production and consumption. Our World in Data. https://ourworldindata.org/energy-production-consumption
66. Rogelj J, Den Elzen M, Höhne N, Fransen T, Fekete H, Winkler H, Schaeffer R, Sha F, Riahi K, Meinshausen M (2016) Paris agreement climate proposals need a boost to keep warming well below 2 c. Nature 534(7609):631–639
67. Romero M, Gonzalez-Aguilar J, Zarza E (2017) Concentrating solar thermal power. In: Energy conversion. CRC Press, pp 655–763
68. Shankar S, Shikha (2017) Renewable and nonrenewable energy resources: bioenergy and biofuels. Principles and applications of environmental biotechnology for a sustainable future, pp 293–314
69. da Silva RC, de Marchi Neto I, Seifert SS (2016) Electricity supply security and the future role of renewable energy sources in brazil. Renew Sustain Energy Rev 59:328–341
70. Singh G (2013) Solar power generation by pv (photovoltaic) technology: a review. Energy 53:1–13. https://doi.org/10.1016/j.energy.2013.02.057
71. Souza Junior MET, Freitas LCG (2022) Power electronics for modern sustainable power systems: Distributed generation, microgrids and smart grids—a review. Sustainability 14(6). https://doi.org/10.3390/su14063597
72. Stambouli AB (2011) Promotion of renewable energies in algeria: strategies and perspectives. Renew Sustain Energy Rev 15(2):1169–1181
73. Tafarte P, Das S, Eichhorn M, Thrän D (2014) Small adaptations, big impacts: options for an optimized mix of variable renewable energy sources. Energy 72:80–92
74. Tiwari G, Tiwari A et al (2016) Handbook of solar energy, vol 498. Springer
75. Trenberth KE (2018) Climate change caused by human activities is happening and it already has major consequences. J Energy Nat Resour Law 36(4):463–481
76. Twidell J (2021) Renewable energy resources. Routledge. https://doi.org/10.4324/9780429452161
77. Uusitalo V, Väisänen S, Inkeri E, Soukka R (2017) Potential for greenhouse gas emission reductions using surplus electricity in hydrogen, methane and methanol production via electrolysis. Energy Convers Manag 134:125–134
78. Woo J, Choi H, Ahn J (2017) Well-to-wheel analysis of greenhouse gas emissions for electric vehicles based on electricity generation mix: a global perspective. Transp Res Part D: Transp Environ 51:340–350
79. Wood AJ, Wollenberg BF, Sheblé GB (2013) Power generation, operation, and control. Wiley
80. Yang P (2024) Hydropower. Springer, pp 109–138. https://doi.org/10.1007/978-3-031-49125-2_4
81. Zou C, Zhao Q, Zhang G, Xiong B (2016) Energy revolution: from a fossil energy era to a new energy era. Nat Gas Ind B 3(1):1–11. https://doi.org/10.1016/j.ngib.2016.02.001

Chapter 2
Multi-criteria Decision-Making

Abstract Multi-Criteria Decision Making (MCDM) methods are techniques that facilitate decision-making in contexts where several criteria need to be evaluated simultaneously. This approach is key in complex situations, as it allows for the analysis of different aspects of a problem, such as technical, economic, social, and environmental factors, in an integrated manner. Within MCDM, there are different methods, with or without the intervention of expert groups, and each one offers a unique procedure for assigning weights and evaluating alternatives. MCDM are applied in various fields related to the optimal choice of alternatives in general. These methods are especially useful in decisions that involve both qualitative and quantitative factors, as they help structure the evaluation process, promoting transparency and consistency in comparisons between options. The ultimate purpose of MCDM is to offer a solid and systematic basis that allows decision-makers to identify the alternative that best fits their objectives and constraints, thus improving results in highly complex scenarios.

2.1 Introduction to MCDM

Multi-Criteria Decision-Making (MCDM) is a branch of operations research that deals with making decisions in the presence of multiple, often conflicting criteria [35]. It provides a structured approach to evaluating various alternatives based on different factors or objectives [20]. MCDM techniques help decision-makers analyse complex problems by breaking them down into manageable components, assessing the importance of each criterion, and synthesizing the information to reach an optimal or satisfactory solution [4].

In the context of energy planning, MCDM is particularly valuable due to the complexity and multi-faceted nature of energy-related decisions [27]. Energy planning often involves balancing various objectives, such as [9]:

- Economic factors (costs, investments, returns).
- Environmental impact (emissions, resource depletion).
- Social considerations (job creation, energy access).

- Technical feasibility (efficiency, reliability).
- Energy security and independence.

Consequently, MCDM techniques can help policymakers, utilities, and other stakeholders in the energy sector to [17]:

- Evaluate different energy sources and technologies.
- Optimize energy mix and resource allocation.
- Assess renewable energy project locations.
- Analyse energy policy options.
- Plan for long-term energy infrastructure development.

By applying MCDM methods, decision-makers can systematically consider multiple criteria, stakeholder preferences, and uncertainties inherent in energy planning. This leads to more transparent, justifiable, and potentially more sustainable energy decisions.

The key components of MCDM techniques explained afterwards include:

A_i Alternatives, where $i = 1, ..., m$.
C_j Criteria, where $j = 1, ..., n$.
x_{ij} Evaluation of alternative A_i with respect to criterion C_j.
w_j Weight of the criterion C_j, usually derived from a weighting phase, where $j = 1, ..., n$.

2.2 Weighting the Criteria

2.2.1 AHP

The Analytic Hierarchy Process (AHP), developed by Thomas Saaty in 1980, is a method designed to resolve MCDM problems by structuring the problem into a hierarchy [30]. AHP incorporates three key principles:

1. Decomposing the decision-making problem into a hierarchy of goals, criteria, and alternatives.
2. Performing pairwise comparisons of elements at each hierarchical level concerning their relationship to the level above.
3. Synthesizing the resulting judgments across different levels of the hierarchy to identify the best alternative.

AHP provides decision-makers with a systematic approach to select the alternative that most effectively aligns with their defined goal [8].

2.2 Weighting the Criteria

2.2.1.1 How Does the AHP Method Work?

The process for applying AHP is as follows [13]:

1. **Problem structuring as a hierarchy**: The decision problem is structured as a hierarchy, as illustrated in Fig. 2.1.

   - At the top of the hierarchy, the main objective is positioned, representing the overarching goal of the decision problem.
   - Below the objective, criteria are established, representing the attributes or factors upon which decisions will be based. These criteria form the foundation of the decision-maker's preferences.
   - At the bottom level are the alternatives, which are the possible options to resolve the decision problem.

2. **Establishing criteria priorities**: A priority vector is calculated to represent the relative importance of each criterion. This is achieved through pairwise comparisons, generating a priority matrix W, where each element reflects the relative importance of the criteria. Comparisons are performed using Saaty's scale (shown in Table 2.1). Experts typically carry out this step, ensuring the judgments reflect informed priorities. To derive the weights, the priority matrix W is solved using Eq. (2.1), which represents the pairwise comparison of criteria:

$$\begin{pmatrix} \frac{w_1}{w_1} & \cdots & \frac{w_1}{w_n} \\ \vdots & \ddots & \vdots \\ \frac{w_n}{w_1} & \cdots & \frac{w_n}{w_n} \end{pmatrix} \begin{pmatrix} w_1 \\ \vdots \\ w_n \end{pmatrix} = \lambda \begin{pmatrix} w_1 \\ \vdots \\ w_n \end{pmatrix}, \qquad (2.1)$$

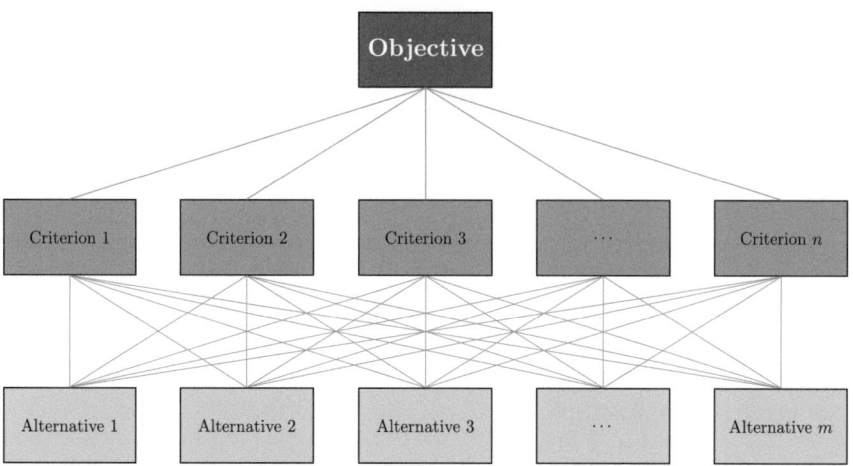

Fig. 2.1 Hierarchy in AHP

Table 2.1 Fundamental pairwise comparison scale proposed by Saaty

Scale	Verbal scale	Explanation
1	Equal importance	Two criteria contribute equally to the objective
3	Moderate importance	Experience and judgement favour one criterion over another
5	Strong importance	One criterion is strongly favoured
7	Very strong importance	One criterion is very dominant
9	Extreme importance	One criterion is favoured by at least one order of magnitude of difference

where W is the matrix, and λ is the eigenvalue corresponding to the priority vector **w**.

By normalizing the priority matrix, the column sums yield the weight vector $\mathbf{w} = [w_1, w_2, \ldots, w_n]$, which represents the relative importance of each criterion. To verify the consistency of the judgments, the consistency index (CI) is calculated using Eq. (2.2):

$$CI = \frac{\lambda_{\max} - n}{n - 1}, \qquad (2.2)$$

where $\lambda_{\max}$ is the principal eigenvalue, and n is the number of criteria. This index is compared to the random consistency index (RI), as defined in Table 2.2 [2]. The consistency ratio (CR) is then calculated as follows:

$$CR = \frac{CI}{RI}. \qquad (2.3)$$

A consistency ratio less than or equal to the thresholds specified in Table 2.3 ensures that the judgments are consistent and acceptable.

3. **Prioritizing alternatives for each criterion**: Once the priorities of the criteria are determined, pairwise comparisons are conducted for the alternatives within each criterion. This follows the same process as for the criteria, using Saaty's scale and consistency checks. The resulting priority vectors are combined into a decision matrix, as in Table 2.4, which reflects the alternatives' relative performance on each criterion.
4. **Aggregating priorities**: The final step involves calculating the overall priority of each alternative. Any of the different ranking of alternatives' techniques can be used. However, Saaty proposed to use the weighted sum method, detailed in Sect. 2.3.1.

2.2 Weighting the Criteria

Table 2.2 Random consistency index (RI) depending on the dimension of the matrix (n)

n	1	2	3	4	5	6	7	8
RI	0	0	0.525	0.882	1.115	1.252	1.341	1.404
n	9	10	11	12	13	14	15	16
RI	1.452	1.484	1.513	1.535	1.555	1.570	1.583	1.595

Table 2.3 CR threshold for n dimension. *Source* Lozano [18]

n	3	4	$n \geq 5$
CR	0.05	0.08	0.10

Table 2.4 Decision matrix

	w_1	w_2	...	w_j	...	w_n
	C_1	C_2	...	C_j	...	C_n
A_1	x_{11}	x_{12}	...	C_{1j}	...	x_{1n}
A_2	x_{21}	x_{22}	...	C_{2j}	...	x_{2n}
:	:	:	:	:	:	:
A_i	x_{i1}	x_{i2}	...	C_{ij}	...	x_{in}
:	:	:	:	:	:	:
A_m	x_{m1}	x_{m2}	...	C_{mj}	...	x_{mn}

2.2.1.2 Advantages and Limitations

The AHP method offers several advantages, including a solid theoretical foundation and strong practical performance. It simplifies complex decision-making by using a hierarchical structure that mirrors the natural way the human mind organizes information. AHP provides a flexible and easily understandable model, allowing for the measurement of intangible factors and prioritization of alternatives. It also synthesizes diverse judgments without requiring consensus and enables iterative refinement of decisions, making it a robust tool for addressing complex problems and balancing trade-offs among competing criteria.

However, AHP has some limitations, such as the need to justify the assumption of independence in hierarchical modelling and the use of a fundamental scale for pairwise comparisons, which can be subjective. The process of prioritization using eigenvector calculations and the method for evaluating the consistency of judgments may also be challenging. Additionally, introducing new alternatives can disrupt the decision-maker's preference structure or highlight inconsistencies in the analysis [23].

The AHP method is a MCDM technique that breaks down a complex problem into a hierarchy of criteria and sub-criteria. It is applied in situations where it is necessary to prioritize or evaluate options in sectors such as management, planning or engineering. Its main advantage is the ability to quantify qualitative criteria through pairwise comparisons, allowing for a more structured and logical evaluation. AHP facilitates decision-making based on expert judgment and detailed priority analysis.

2.2.2 Entropy

The Entropy method is a widely used tool in MCDM. The main goal of the method is to provide an objective measure of the degree of dispersion or uncertainty present in the data related to the different evaluated criteria. It is based on the concept of entropy in information theory, proposed by Claude Shannon in 1948 [32], which measures the degree of disorder or uncertainty in a given system.

In the context of MCDM, Entropy allows calculating the objective weights of the criteria without direct intervention from an expert or decision-maker, unlike other methods such as the AHP (refer to Sect. 2.2.1) or the Analytic Network Process (ANP). This makes it an objective and bias-free option for weighting criteria when detailed information on their relative importance is not available [21].

2.2.2.1 How Does the Entropy Method Work?

The Entropy method in MCDM follows a series of steps to calculate the objective weights of the criteria, based on the evaluation data of the alternatives [28]. The general process is described below:

1. **Normalized decision matrix**: The first step is to normalize the decision matrix to ensure that all variables are on the same scale and are comparable. The normalization is performed as follows:

$$r_{ij} = \frac{x_{ij}}{\sum_{i=1}^{m} x_{ij}}, \qquad (2.4)$$

where r_{ij} is the normalized value of alternative i under criterion j. This process transforms the values so that the sum of the alternatives under each criterion equals 1.

2.2 Weighting the Criteria

2. **Entropy calculation**: Once the matrix is normalized, the entropy Ent_j for each criterion is calculated using the following expression:

$$Ent_j = -k \sum_{i=1}^{m} r_{ij} \ln(r_{ij}), \qquad (2.5)$$

where $k = \frac{1}{\ln(m)}$, which ensures that the entropy is normalized between 0 and 1. If all r_{ij} values are equal, the entropy is maximized, indicating high uncertainty or low discriminative power of the criterion. On the other hand, when the values are highly dispersed, the entropy is low, suggesting that the criterion has a greater discriminative power.

3. **Dispersion degree calculation**: For each criterion, its degree of dispersion dd_j is calculated as the complement of its entropy:

$$dd_j = 1 - Ent_j. \qquad (2.6)$$

This value reflects how informative the criterion is: a value close to 1 indicates that the criterion has great capacity to distinguish between alternatives, while a value close to 0 implies that the criterion does not provide sufficient information.

4. **Determining the criteria weights**: Finally, the weights w_j of the criteria are calculated from their degrees of dispersion. The weight of each criterion is determined by dividing its degree of dispersion by the sum of the degrees of dispersion of all criteria:

$$w_j = \frac{dd_j}{\sum_{j=1}^{n} dd_j}. \qquad (2.7)$$

In this way, the assigned weights reflect the ability of each criterion to discriminate among the alternatives. Criteria with greater discriminative power receive higher weights.

2.2.2.2 Advantages and Limitations

The Entropy method has found applications in various fields, such as sustainability evaluation, project selection, risk management, and decision-making in engineering and finance [7, 36]. Some of its key advantages include:

- **Objectivity**: As it does not depend on human intervention for weight assignment, the method is more objective compared to other methods that require subjective judgments.
- **Ability to handle large datasets**: It is suitable for situations where large amounts of information are handled, as its algorithmic approach allows working with multiple criteria and alternatives without added complexity.

- **Ease of integration**: It can be easily combined with other MCDM methods, such as the TOPSIS (refer to Sect. 2.3.2) or the Preference Ranking Organization Method for Enrichment of Evaluations (PROMETHEE).

Despite its advantages, the entropy method also presents some limitations:

- **Dependence on data quality**: The method assumes that the input data is accurate and complete. If there are errors or inconsistencies in the data, the results may not be reliable.
- **Inability to capture subjective preferences**: Since it is purely objective, the method does not account for preferences or expert knowledge, which can be relevant in certain applications where decision-makers have a better understanding of the problem.
- **Ambiguity in entropy interpretation**: While entropy can indicate the amount of uncertainty associated with each criterion, interpreting the practical meaning of this value can be complex in certain contexts.

> The entropy method in MCDM is a powerful tool for assigning objective weights to criteria when subjective information is not available or when bias is to be reduced. Its applicability in a wide variety of fields and its mathematical approach make it an attractive option for multi-criteria decision-making, especially when a more objective analysis of data is sought. However, decision-makers should be mindful of the inherent limitations of this approach, particularly in situations where subjective preferences play a significant role.

2.2.3 Compromised AHP + Entropy

The combination of the AHP and the Entropy methods in MCDM is referred to as the Compromised AHP + Entropy method. This hybrid approach harnesses the strengths of both methods to enhance decision-making, particularly in cases where both subjective judgments and objective data need to be integrated [6]. As the AHP method allows decision-makers to assign subjective weights through pairwise comparisons (which can introduce inconsistencies or biases), and the Entropy method objectively assigns weights based on the distribution of data (avoiding human intervention), this hybrid approach mitigates the limitations of each method. Consequently, the integration of these two methods creates a more robust decision-making process, ensuring that both subjective preferences and objective data are considered.

2.2.3.1 How Does the Compromised AHP + Entropy Method Work?

In the Compromised AHP + Entropy method, the subjective weights derived from AHP and the objective weights calculated from the Entropy method are combined to balance the subjective and objective information [14]. The process is as follows:

1. **AHP weight calculation**: Pairwise comparisons are used to compute the subjective weights of the criteria, reflecting the preferences of the decision-makers (refer to steps 1–2 of Sect. 2.2.1).
2. **Entropy weight calculation**: The Entropy method is applied to calculate the objective weights based on the variability of the data across the criteria (refer to Sect. 2.2.2).
3. **Weight integration**: The final weight for each criterion is derived by combining the subjective and objective weights. A common approach is to use a weighted sum:

$$w_{\text{final}} = \alpha \cdot w_{\text{AHP}} + (1 - \alpha) \cdot w_{\text{Entropy}}, \tag{2.8}$$

where α reflects the relative importance of subjective preferences (AHP) and $1 - \alpha$ represents the weight given to the objective data (Entropy).

2.2.3.2 Advantages and Limitations

This hybrid approach provides several advantages in MCDM:

- **Balanced decision-making**: By integrating subjective and objective weights, the method offers a more balanced decision-making process.
- **Reduction of bias**: The inclusion of objective data reduces the influence of personal bias inherent in the AHP method.
- **Flexibility**: The method can be applied in a wide variety of decision-making scenarios, especially those involving both quantitative data and expert judgment.

While powerful, this method has some limitations:

- **Complexity**: Combining subjective and objective data increases the computational and cognitive effort required.
- **Dependence on data quality**: The accuracy of the results depends heavily on the quality and completeness of the data.
- **Subjectivity in Integration**: The choice of α when combining AHP and Entropy weights introduces subjectivity, and different values may yield different results.

> The Compromised AHP + Entropy method is a versatile tool in MCDM, offering a robust framework for decision-making that integrates both subjective preferences and objective data. Its application across various domains, from renewable energy to project management, highlights its flexibility and utility. However, decision-makers should be aware of the complexity and data requirements when applying this hybrid approach.

2.3 Ranking of Alternatives

2.3.1 Weighted Sum Method

This is the simplest technique to rank the different alternatives on a MCDM problem. It is done by aggregating the priority values from the valuation matrix with the weight vector **w** obtained for the criteria. The global priority vector **p** of the alternatives is computed using:

$$p_i = \sum_{j=1}^{n} w_j \cdot x_{ij}, \tag{2.9}$$

where w_j represents the weight for each criterion, and x_{ij} is the performance of alternative A_i on criterion C_j. The best alternative is the one with the highest value in the global priority vector.

2.3.2 TOPSIS

The Technique for Order of Preference by Similarity to Ideal Solution (TOPSIS) was originally developed by Ching-Lai Hwang and Yoon in 1981 with further developments by Yoon in 1987, and Hwang, Lai and Liu in 1993 [33]. It is based on the principles of ideal and anti-ideal solutions for the selection of alternatives. The best alternative minimizes the distance to the ideal solution while maximizing the distance to the anti-ideal solution [26].

Figure 2.2 illustrates the method with five alternatives ($A_1, \ldots, A_5$), two criteria (C_1 and C_2), and the ideal and anti-ideal points. In this case, A_3 is the closest alternative to the ideal solution, while A_2 and A_4 are the furthest from the anti-ideal solution. TOPSIS uses weighted distances to compute the proximity to both ideal and anti-ideal solutions, relying on multivariate data analysis [29].

2.3 Ranking of Alternatives

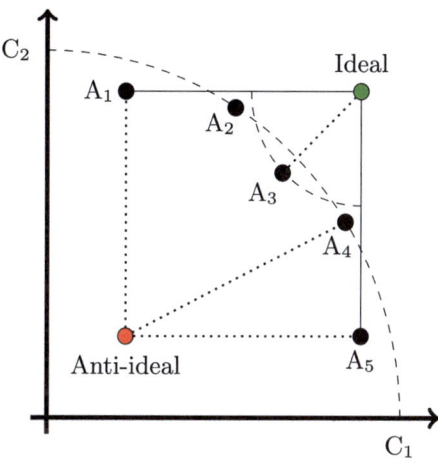

Fig. 2.2 Ideal and anti-ideal alternatives in the TOPSIS method

2.3.2.1 How Does the TOPSIS Method Work?

The TOPSIS procedure is detailed in [12] and consists of the following steps:

1. **Construct the decision matrix**: The alternatives $A_1, A_2, \ldots, A_m$ are evaluated based on n criteria $C_1, C_2, \ldots, C_n$, forming a decision matrix, such as the one in Table 2.4. Each entry x_{ij} represents the evaluation of alternative A_i against criterion C_j. The weight vector $\mathbf{w} = [w_1, w_2, \ldots, w_n]$ assigns importance to each criterion, with weights obtained from a different method (such as the AHP, presented in Sect. 2.2.1), satisfying $\sum_{j=1}^{n} w_j = 1$.

2. **Normalize the decision matrix**: Normalize the values from Table 2.4 using the following formula:
$$\overline{n}_{ij} = \frac{x_{ij}}{\sqrt{\sum_{i=1}^{m} x_{ij}^2}}. \tag{2.10}$$

3. **Generate the weighted normalized matrix**: The normalized matrix is weighted according to the criteria weights:
$$\overline{v}_{ij} = w_j \cdot \overline{n}_{ij}. \tag{2.11}$$

4. **Identify the ideal and anti-ideal solutions**: The ideal solution $\overline{A}^+$ and anti-ideal solution $\overline{A}^-$ are computed as:
$$\overline{A}^+ = \{\overline{v}_1^+, \ldots, \overline{v}_n^+\} = \left\{(\max_i \overline{v}_{ij}, j \in J), (\min_i \overline{v}_{ij}, j \in J')\right\} \tag{2.12}$$

$$\overline{A}^- = \{\overline{v_1^-}, \ldots, \overline{v_n^-}\} = \left\{ (\min_i \overline{v}_{ij}, j \in J), (\max_i \overline{v}_{ij}, j \in J') \right\}, \quad (2.13)$$

where J refers to maximizing criteria and J' refers to minimizing criteria.

5. **Calculate the distance to the ideal solutions**: The Euclidean distance from each alternative to the ideal and anti-ideal solutions is computed as follows:

- Distance to the ideal solution $\overline{d}_i^+$:

$$\overline{d}_i^+ = \sqrt{\sum_{j=1}^{n} (\overline{v}_{ij} - \overline{v}_j^+)^2}. \quad (2.14)$$

- Distance to the anti-ideal solution $\overline{d}_i^-$:

$$\overline{d}_i^- = \sqrt{\sum_{j=1}^{n} (\overline{v}_{ij} - \overline{v}_j^-)^2}. \quad (2.15)$$

6. **Compute the relative proximity to the ideal solution**: The relative proximity $\overline{R}_i$ of each alternative to the ideal solution is given by:

$$\overline{R}_i = \frac{\overline{d}_i^-}{\overline{d}_i^+ + \overline{d}_i^-}. \quad (2.16)$$

7. **Rank the alternatives**: Based on the values of $\overline{R}_i$, the alternatives are ranked. The higher the value of $\overline{R}_i$ (closer to 1), the better the alternative.

2.3.2.2 Advantages and Limitations

The TOPSIS method offers several benefits. First, it is highly flexible, handling any number of evaluation objects and indicators in the data without restrictions. Regardless of the size or nature of the dataset, TOPSIS identifies both the "positive ideal solution" and the "negative ideal solution" and calculates the relative closeness of each object to these solutions. The method is data-driven and unaffected by the subjective judgments of decision-makers, resulting in more objective evaluations. Additionally, TOPSIS provides clear rankings of evaluation objects, making it easier for decision-makers to reach conclusions. Furthermore, it fully utilizes the original data, allowing for a holistic and detailed analysis of the results.

Despite its advantages, the TOPSIS method has limitations. The construction of the "positive ideal solution" by combining the best indicators and the "negative ideal solution" by combining the worst, can lead to unrealistic results if the indicators conflict. This means the solutions identified may not reflect the real-world situation, necessitating careful consideration of the context and criteria during evaluation, which increases the workload [3].

Additionally, TOPSIS ranks alternatives based on relative closeness, which may not provide the same level of clarity as the weighting results from other methods like AHP or entropy. The numerical ranking of closeness does not always allow decision-makers to distinguish specific grades or categories for each alternative. Finally, the method is sensitive to the weights assigned to the indicators, which can introduce subjectivity into the evaluation. Variations in weight settings can significantly affect the final outcome, potentially impacting the objectivity of the results [5].

> The TOPSIS method is a MCDM tool that seeks to identify the best alternative by comparing it with a positive and a negative ideal solution. Its applicability is broad, being useful in selection or classification problems in areas such as engineering, economics or management. One of its main advantages is its simplicity and ability to handle multiple criteria, allowing alternatives to be prioritized effectively and objectively.

2.3.3 VIKOR

The VlseKriterijumska Optimizacija I Kompromisno Resenje (VIKOR), which in Serbian means "multi-criteria optimization and compromise solution" method, is a popular MCDM tool. Developed by Opricovic in 1998 and refined in 2004 [25], VIKOR is designed to solve decision problems where the decision-maker must balance conflicting criteria, such as cost and quality, or risk and benefit.

The VIKOR method is grounded in the concept of identifying a "compromise solution", which is defined as a solution that is acceptable to most stakeholders. VIKOR distinguishes itself from other MCDM methods by focusing on providing a solution that balances conflicting criteria, assessing how far each alternative is from an ideal solution, and allowing for negotiation between decision-makers, if necessary.

2.3.3.1 How Does the VIKOR Method Work?

The VIKOR method follows several steps aimed at finding a solution close to the ideal compromise, optimizing each criterion without drastically sacrificing others. The main steps are as follows [1]:

1. **Normalize the data**: The data are normalized to ensure that all criteria can be compared fairly, as they may be measured in different units. Normalization is performed using the following expression:

$$x_{ij}^n = \frac{x_{ij} - x_j^{min}}{x_j^{max} - x_j^{min}} \quad \text{for maximizing criteria}, \tag{2.17}$$

$$x_{ij}^n = \frac{x_j^{max} - x_{ij}}{x_j^{max} - x_j^{min}} \quad \text{for minimizing criteria}, \tag{2.18}$$

where x_{ij} represents the original value of alternative i for criterion j, and x_j^{max} and x_j^{min} are the maximum and minimum values observed across all alternatives for criterion j, respectively.

2. **Determine the ideal and anti-ideal solutions**: Once the data are normalized, the ideal solution A^+ is defined as the alternative that maximizes benefits and minimizes costs across all criteria, i.e., the one with the best possible performance values for each criterion. Analogously, the anti-ideal solution A^- is determined, representing the worst values for each criterion.
3. **Calculate the S_i, R_i, and Q_i indices**: Three indices are calculated for each alternative:

   - S_i index: Represents the weighted sum of the distances of alternative i from the ideal solution A^+, calculated as:

   $$S_i = \sum_{j=1}^{n} w_j \cdot \frac{f_j^+ - f_{ij}}{f_j^+ - f_j^-}, \tag{2.19}$$

   where w_j is the weight of criterion j, and f_j^+ and f_j^- are the ideal and anti-ideal values for criterion j, respectively.
   - R_i index: Represents the maximum distance to the ideal solution for each criterion, reflecting the worst difference:

   $$R_i = \max_j \left(w_j \cdot \frac{f_j^+ - f_{ij}}{f_j^+ - f_j^-} \right). \tag{2.20}$$

   - Q_i index: Combines S_i and R_i to obtain a compromise value balancing total and maximum distances, using a balance parameter v (typically, $v = 0.5$):

   $$Q_i = v \cdot \frac{S_i - S^+}{S^- - S^+} + (1 - v) \cdot \frac{R_i - R^+}{R^- - R^+}, \tag{2.21}$$

   where S^+, R^+, S^-, and R^- represent the minimum and maximum values of S_i and R_i across all alternatives.

4. **Ranking and selecting the best alternative**: Finally, the alternatives are ranked based on their Q_i, S_i, and R_i indices. The alternative with the lowest Q_i value is considered the best compromised solution. To be accepted, this alternative must meet two conditions:

2.3 Ranking of Alternatives

- *Acceptance condition*: The alternative must be sufficiently close to the second-best solution, indicating no significant ranking differences.
- *Consensus condition*: If there is no consensus, the method suggests further negotiation or analysis to reach an acceptable conclusion.

2.3.3.2 Advantages and Limitations

The VIKOR method is particularly useful in situations where decision-makers seek a compromise solution rather than one that optimizes a single criterion. Some of its key advantages include:

- **Flexibility**: It handles conflicting criteria and provides a compromise solution, which is critical in multi-interest scenarios.
- **Clarity**: The S_i, R_i, and Q_i indices offer clear and transparent rankings of alternatives, making the decision process straightforward.

> The VIKOR method is a powerful tool for MCDM that helps decision-makers find compromise solutions in problems with conflicting criteria. Its flexibility and structured approach make it a popular choice across various applications. Additionally, its ability to facilitate consensus among stakeholders makes it valuable in scenarios where single-criterion optimization is insufficient.

2.3.4 SIMUS

The Sequential Interactive Modeling for Urban Systems (SIMUS) method is an advanced mathematical model based on linear programming. Initially developed to address urban planning issues, its application has expanded to various areas, including energy projects, environmental management, construction, and economics, among other fields requiring a multi-criteria approach for decision-making [11, 15, 34]. This method was developed by Nolberto Munier [22], a researcher at the Polytechnic University of Valencia, Spain.

2.3.4.1 How Does the SIMUS Method Work?

SIMUS is based on solving multiple scenarios expressed as linear programs. The process begins with the construction of a decision matrix, which contains the alternatives or projects and the criteria or factors used for their evaluation. The linear program is then executed using a selected criterion (factor) as the objective to be optimized in the first iteration. The step-by-step process is as follows:

1. **Optimization for each criterion**: A criterion is chosen as the objective in the linear program, and the problem is solved to obtain optimal values. These results are stored in a new matrix called the Efficient Results Matrix (ERM).
2. **Evaluating all factors**: The process is repeated for each criterion until all have been evaluated. The results from each iteration are added to the ERM, which reflects the performance of each alternative with respect to each criterion.
3. **Analysis and classification**: Once the ERM is completed, two types of classifications for the alternatives are generated:

   - *ERM-based classification*: This analysis is conducted vertically across the ERM. It assesses the values of each alternative concerning all objectives. A score is assigned to each alternative by summing the values of all the solutions it participates in. This score is adjusted by a coefficient based on how often each alternative appears in an optimal solution.
   - *Project Dominance Matrix (PDM)*: This analysis is conducted horizontally, evaluating how many times an alternative ranks higher than others or is outperformed by others. This determines which alternatives are dominant within the set of proposed projects or solutions.

Based on the analysis of both classifications (ERM and PDM), three possible outcomes arise:

- **Match between classifications**: If both vertical and horizontal classifications match, the result is reliable and positive, making the decision-making process clear.
- **Minor differences**: If there are slight discrepancies between both classifications, these can provide additional insights and help refine the decision.
- **Significant differences**: If the classifications are completely different, a review of the input data and the weight assignments to the criteria is necessary, as there may be inconsistencies in the model or the initial preferences.

One of the method's key strengths is its ability to customize data and adapt to the decision maker's needs. The matrix can handle different data types, such as decimal, binary, or integer values, which enhances its flexibility in diverse contexts. Additionally, users can introduce custom weights for each criterion, allowing them to adjust the relative importance of the factors in the decision-making process.

2.3.4.2 Advantages and Limitations

The SIMUS method is highly flexible and adaptable, making it suitable for a variety of decision-making problems. Its ability to incorporate both qualitative and quantitative criteria makes it especially useful in complex scenarios where multiple factors must be considered. As a MCDM method, SIMUS allows the evaluation of alternatives based on a range of factors simultaneously, which is essential for resource allocation or spatial analysis problems. Additionally, the interactive nature of the method enables decision-makers to iteratively adjust weights, objectives, and constraints as needed, providing a modelling experience that can adapt to changes in criteria or the problem's context. Another significant advantage is the clarity in interpreting results: SIMUS provides a ranking of alternatives that highlights the relative advantages of each option based on each criterion, helping decision-makers understand the benefits of each alternative. Moreover, SIMUS effectively handles conflicting objectives, such as cost minimization versus quality maximization, by calculating trade-offs and offering balanced solutions.

However, SIMUS has some limitations. For large-scale or highly complex problems, the computational demands of the method may increase, making implementation and interpretation difficult without solid computational resources. It also relies on accurate weighting of criteria, which requires careful and often subjective input from experts, introducing the risk of biases in the analysis. The choice and definition of criteria can significantly impact the results, as SIMUS is sensitive to the selected criteria, meaning that the conclusions might vary substantially if the criteria are not well-defined or suitable for the context. While the interactive process is advantageous, it can also be time-consuming, particularly if multiple rounds of adjustments are needed to achieve an optimal solution. Finally, SIMUS requires a set of detailed and precise data inputs for each criterion, and in contexts where data is scarce or difficult to obtain, its applicability may be limited.

> The SIMUS method is a multi-criteria decision-making and optimization technique based on linear programming. It is mainly applied in resource optimization and planning in areas such as logistics, project management, and urban planning. Its main advantage is that it allows for handling large amounts of data and constraints, generating optimal solutions for multiple objectives simultaneously. In addition, SIMUS facilitates an interactive process to refine and adjust the results based on the decision-maker's priorities.

2.4 Fuzzy

Fuzzy logic is a method based on heuristic rules of the form "If-Then" to relate contextualized random values. For this purpose, common linguistic terms such as "very much", "a little" or "worse" are used. The theory of fuzzy sets was formulated by Zadeh [38]. In this theory, operations between fuzzy sets (union, intersection, difference,...), and other operations on sets based on fuzzy logic are defined.

2.4.1 Fuzzy Sets

Fuzzy set theory establishes a definition of the data using simple language such as lowest risk, high probability or moderate impact. These terms are clearer and more intuitive to use than mathematical values. Fuzzy set theory establishes that, for each fuzzy set, there is an associated membership function for its elements, which indicates the extent to which the element is part of that fuzzy set. Typical forms of membership functions are trapezoidal, triangular, linear, and curved.

According to fuzzy sets theory [10, 16], on a universal set X, a fuzzy subset A of X is defined by a membership function $A(x)$, which assigns each element x in X to a real number in the interval [0,1]. Following this, the value of the function $\mu_A(x)$ is the level of membership of x in A. When $\mu_A(x)$ is large, the degree of membership of x in A is strong.

The main membership functions, or fuzzy numbers, used in cases with high uncertainty are triangular fuzzy numbers (Fig. 2.3) and trapezoidal fuzzy numbers (Fig. 2.4).

- **Triangular fuzzy number**: A triangular fuzzy number is identified as $A = (a, b, c)$, where a, b, and c are real numbers $\mathbb{R}$, being $a < b < c$.

  A triangular fuzzy number is also defined as $A = x, \mu_A(x)$, where x is the value in the interval R and its membership function, $\mu_A : R \to [0, 1]$, has the following properties:

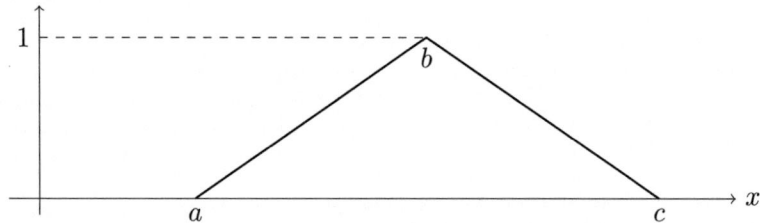

Fig. 2.3 Triangular fuzzy number

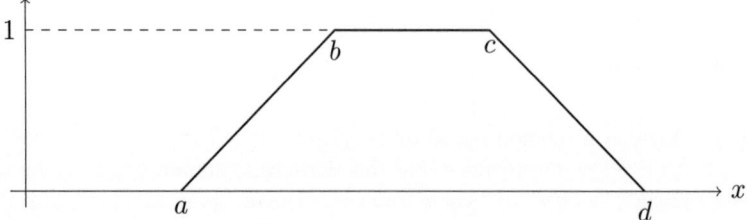

Fig. 2.4 Trapezoidal fuzzy number

2.4 Fuzzy

- Constant in $(-\infty, a]$: $\mu_A(x) = 0$, $\forall x \in (-\infty, a]$,
- Strictly increasing in $[a, b]$,
- Strictly decreasing in $[b, c]$,
- Constant in $[c, \infty)$: $\mu_A(x) = 0$, $\forall x \in [c, \infty)$.

If μ_A^L is defined as the left-hand side of the membership function of a fuzzy number A, defining $\mu_A^L(x) = \mu_A(x)$, $\forall x \in [a, b]$, and μ_A^R as the right-hand side of the membership function of the fuzzy number A, defining $\mu_A^R(x) = \mu_A(x)$, $\forall x \in [b, c]$, its membership function can be defined as:

$$\mu_A(x) = \begin{pmatrix} 0, \forall x < a \\ \mu_A^L(x) = \dfrac{x-a}{b-a}, \forall a \leq x \leq b \\ 1, \forall x = b \\ \mu_A^L(x) = \dfrac{x-c}{b-c}, \forall b \leq x \leq c \\ 0, \forall x > c \end{pmatrix} \quad (2.22)$$

- **Trapezoidal fuzzy number**: A trapezoidal fuzzy number is identified as $A = (a, b, c, d)$, where a, b, c, and d are real numbers $\mathbb{R}$, being $a < b < c < d$.

A trapezoidal fuzzy number is also defined as $A = x$, $\mu_A(x)$, where x is the value in the interval R and its membership function, $\mu_A : R \rightarrow [0, 1]$, has the following properties:

- Constant in $(-\infty, a]$: $\mu_A(x) = 0$, $\forall x \in (-\infty, a]$,
- Strictly increasing in $[a, b]$,
- Constant in $[b, c]$,
- Strictly decreasing in $[c, d]$,
- Constant in $[d, \infty)$: $\mu_A(x) = 0$, $\forall x \in [d, \infty)$.

If μ_A^L is defined as the left-hand side of the membership function of a fuzzy number A, defining $\mu_A^L(x) = \mu_A(x)$, $\forall x \in [a, b]$, and μ_A^R as the right-hand side of the membership function of a fuzzy number A, defining $\mu_A^R(x) = \mu_A(x)$, $\forall x \in [c, d]$, its membership function can be defined as:

$$\mu_A(x) = \begin{pmatrix} 0, \forall x < a \\ \mu_A^L(x) = \dfrac{x-a}{b-a}, \forall a \leq x \leq b \\ 1, \forall b \leq x \leq c \\ \mu_A^L(x) = \dfrac{X-d}{c-d}, \forall c \leq x \leq d \\ 0, \forall x > d \end{pmatrix} \quad (2.23)$$

The key difference between selecting a triangular or trapezoidal fuzzy number lies in the experts' level of knowledge consulted. As they express their opinions using linguistic terms, greater knowledge and experience result in setting $b = c$ in a trapezoidal number, effectively transforming it into a triangular number [37].

2.4.2 Uncertainty: Contribution of Fuzzy Logic to Decision-Making

Decision-making involves an inherent analysis of the particular situation under analysis: the variables involved, the known data, and their precision, are some of the elements that must be taken into account when carrying out the decision-making process. In a specific study, a wide range of possibilities, perturbations, changes, complexities, non-linearities, chaos, and evolution must be considered in the analysis. In other words: decision-making takes place in a context dominated by uncertainty and subjectivity.

Fuzzy logic is widely used in decision-making because, by relying on the use of linguistic terms, it allows uncertainty, ambiguity, and imprecision to be managed in a flexible and intuitive way. Likewise, by transforming linguistic opinions into numbers and operating with them, the subjectivity of the analysis is reduced. Therefore, the main application of fuzzy logic is to mitigate the subjectivity of expert opinions in the application of the AHP method, which makes it the ideal complement for this method.

2.4.3 Application Example: Software @Risk

The main disadvantage of classical methods of analysis based on fuzzy logic is the complexity of their resolution and their high computational cost. Classical fuzzy methods share the same general structure [19]:

1. **Parameter definition and measurement**: In this phase, the basic parameters of the analysis are defined. Parameter measurement is carried out using linguistic terms and these are converted into fuzzy numbers. If the group of experts has extensive knowledge of the sector, then they can give concise opinions using linguistic terms, which are then converted into triangular numbers, as already justified in Sect. 2.4.1 [37].
2. **Fuzzy inference phase**: The relationship between the input and output parameters can be defined as "If-Then" relationships, or in the form of mathematical functions defined by an appropriated fuzzy arithmetic operator.
3. **Defuzzification**: As a result of the previous phase, a fuzzy number is obtained which, in this phase, is transformed into an exact numerical value.

The difficulty of the classical methodology lies in the mathematics of the methodology itself. Mathematical operations between two fuzzy numbers return a fuzzy number of the same form, i.e., a mathematical operation between two triangular fuzzy numbers returns a triangular fuzzy number.

To increase the precision of mathematical operations between fuzzy numbers, one of the most traditionally extended methods for phase 2 of the fuzzy inference is the use of $\alpha - cut$, as described in Nieto-Morote and Ruz-Vila [24]. The $\alpha - cut$

2.4 Fuzzy

method is based on cutting the fuzzy number into levels, the greater the number of levels, the greater the precision in the operations. The computational cost increases with the number of levels, and there will always be a loss of data, as it is not possible to make infinite cuts.

If a computational tool is available to handle triangular or trapezoidal numbers as distribution functions, the data loss associated with the $\alpha - cut$ method disappears. Additionally, when this tool operates in a standard software environment like Excel, applying the AHP method for decision-making becomes straightforward and accessible to most teams. The @Risk tool meets both of these criteria.

To test the application of the @Risk tool in the resolution of the AHP method combined with fuzzy logic, Serrano-Gomez and Munoz-Hernandez [31] applied these methodology to qualitative risk analysis in RES construction projects, specifically, in a 250 MW PV power plant project located in Southeastern Spain. A risk assessment group composed of 12 experts with extensive experience in PV construction projects was created.

Once the risks were identified (Table 2.5), questionnaires were completed by the experts to express their opinion on the risk impact (RI) on the scope, cost, and time of the project, as well as the risk probability (RP) of occurrence of the identified risks. An example of how the experts gave their opinion regarding some risks is shown in Table 2.6, according to Figs. 2.5 and 2.6.

As the risk were prioritized using the AHP method (see Sect. 2.2.1), the experts also had to discriminate between risks by means of a pair-wise comparison in terms of the RI on the scope, cost, and time, as well as on the RP. Once the experts' opinions were collected, their opinions were checked for consistency and level of confidence, and the aggregation process is started. As a result of the aggregation process, the Overall Risk Factor (ORF) is determined for each risk. The ORF was calculated for each evaluated risk by using:

$$ORF = \frac{RI}{RDI} \cdot \frac{RP}{RDP} \quad (2.24)$$

where ORF is the overall risk factor function, RDI is risk discrimination impact, and RDP is risk discrimination probability in the project life cycle.

As a result of the aggregation process, the local ORF parameters on the scope, cost, and time are estimated, as well as the global ORF, as shown for some risks in Table 2.7.

After carrying out a Monte Carlo simulation with 100 simulations and 10,000 iterations for each simulation, with the @Risk tool, there are no cases in which the best-fit function is a triangular function, as would result if the $\alpha - cut$ method had been used.

Once applied the model to the case study, a risk ranking has been obtained that allows planning the overall risk response to Scope, Costs, and Time. Based on the ranking results within the general framework project, shown in Fig. 2.7, it will be necessary to focus the efforts of the project management team towards the risks 'Bank financing', 'Changes in energy prices', 'Specific legislation changes', 'The

Table 2.5 Risk list

Risk	
1.1.1	Level of political stability
1.1.2	The change in energy policy
1.2.1	Approval by the Local Body
1.2.2	Obtaining the construction license
2.1.1	Technological climate change adequacy
2.1.2	Flood and storm risks
2.1.3	Estimation of effective solar radiation
2.1.4	Earthworks
2.1.5	Geotechnical study
2.2.1	New PV solar power systems
2.2.2	PV cell selection
2.2.3	Inverters selection
2.2.4	Selection of support panel structure
2.2.5	Connection to the electric grid
2.2.6	Alternative power generation systems
3.1.1	Plant operation cost
3.1.2	Corrective maintenance costs
3.1.3	Prevention of maintenance costs
3.1.4	Performance losses
3.2.1	Errors in estimating the effective solar radiation energy
3.2.2	Revenue estimation due to the climate change
3.2.3	Earthworks resources
3.2.4	Flood prevention works
3.2.5	Solution of geotechnical problems
3.3.1	Connection to electric grid costs
3.3.2	Agreement costs with landowners
3.3.3	Possibility of constructing the power connection infrastructure
3.3.4	Construction license
3.4.1	Costs due to inadequate PV cell selection
3.4.2	Costs due to inadequate inverter selection
3.4.3	Costs due to lack of consistency in the selection of support panels
3.5.1	Bank financing
3.5.2	Changes in power demand
3.5.3	Inflation
3.5.4	Changes in energy prices
4.1.1	Delays in obtaining administrative approval for the connection infrastructure
4.1.2	Construction delays of the power connection infrastructure
4.1.3	Delays in obtaining PV plant Start-up Act
4.1.4	Delays in the agreement signature with REE and CNMC
4.2.1	Delays in obtaining the Local Body Approval

(continued)

Table 2.5 (continued)

4.2.2	Delays in obtaining approval of the environmental impact
4.2.3	Delays in obtaining the construction license
5.1.1	Specific legislation changes
5.1.2	General legislation changes
5.2.1	Legislative changes in the Administrative Authorization of the power connection infrastructure
5.2.2	Legislative changes in the Startup Act permits
5.2.3	Obtaining the electrical registration for production facilities
5.3.1	Legislative changes in the Local Body Approval
5.3.2	Legislative changes in the Environmental Impact Approval
5.3.3	Legislative changes in the Construction License
6.1.1	Theft
6.1.2	Vandalism
6.1.3	Terrorism
6.2.1	Social consequences resulting from land acquisition
6.2.2	Social acceptance

Table 2.6 RI and RP expert opinions

Risk	RI-Scope	RI-Costs	RI-Time	RP
2.1.1	Minor	Moderate	Negligible	Very low
2.1.2	Negligible	Serious	Moderate	High
2.1.3	Serious	Critical	Minor	Low
2.1.4	Negligible	Moderate	Serious	Low
2.1.5	Minor	Serious	Critical	Low

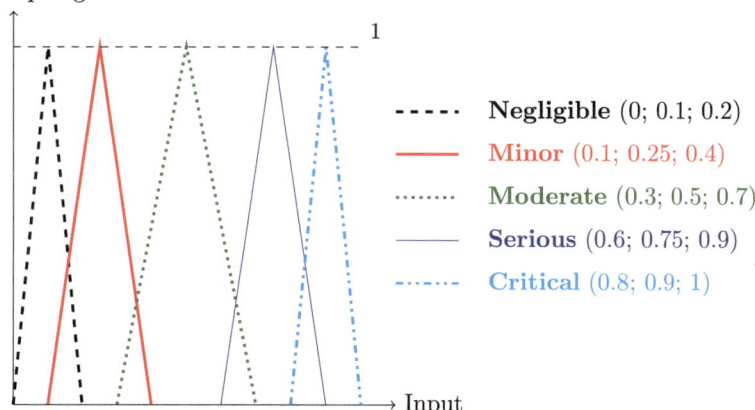

Fig. 2.5 RI linguistic terms and fuzzy numbers

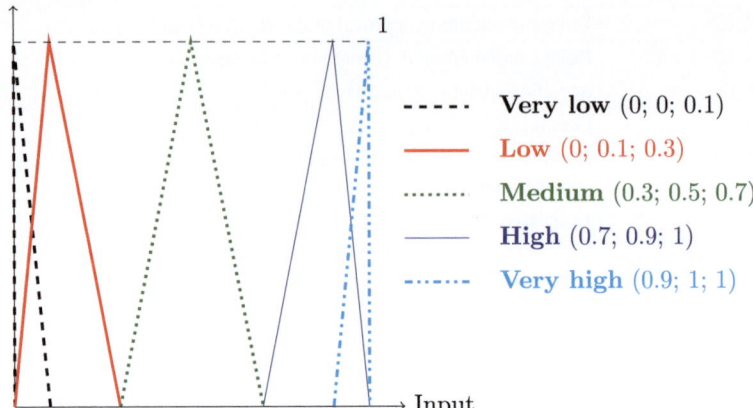

Fig. 2.6 RP linguistic terms and fuzzy numbers

Table 2.7 Estimate ORF: plant location

Risk	ORF Schedule	ORF Costs	ORF Time	ORF
2.1.1	Invgauss (7.31; 87.71; Shift(1.74))	Lognorm (5.54; 1.53; Shift(1.26))	Gamma (6.71; 0.55; Shift(2.50))	Invgauss (58.24; 244.67; Shift(4.69))
2.1.2	Gamma (5,53;1,43; Shift(6,95))	Gamma (5,21;1,94; Shift(9,55))	Invgauss (12,87;127,43; Shift(4,65))	Invgauss (245,34;913,71; Shift(29,32))
2.1.3	Gamma (5,49;2,60; Shift(12,30))	Gamma (4,63;3,03; Shift(13,83))	Gamma (5,99;1,04; Shift(4,99))	Invgauss (544,80;2028,92; Shift(64,71))
2.1.4	Gamma (5,77;0,68; Shift(3,44))	Gamma (4,67;1,02; Shift(4,90))	Gamma (5,18;1,23; Shift(5,82))	Gamma (2,38;36,58; Shift(39,44))
2.1.5	Gamma (5,17;1,43; Shift(7,56))	Gamma (5,08;1,43; Shift(7,40))	Gamma (5,24;1,97; Shift(9,96))	Invgauss (321,99;1245,72; Shift(49,12))

change in energy policy', 'Delays in obtaining the construction licence'. Of the ten most influential risks in the general framework of the project, the majority are those due to delays. Therefore, in order to ensure the future success of the project, the risk management team should focus its efforts on controlling project deadlines, whilst not forgetting bank financing and possible changes in the electricity market price.

In order to compare the operation of the proposed model, the risk ranking has been determined using the classical fuzzy methodology ($\alpha - cut$), resulting in the ranking shown in Fig. 2.8.

2.4 Fuzzy

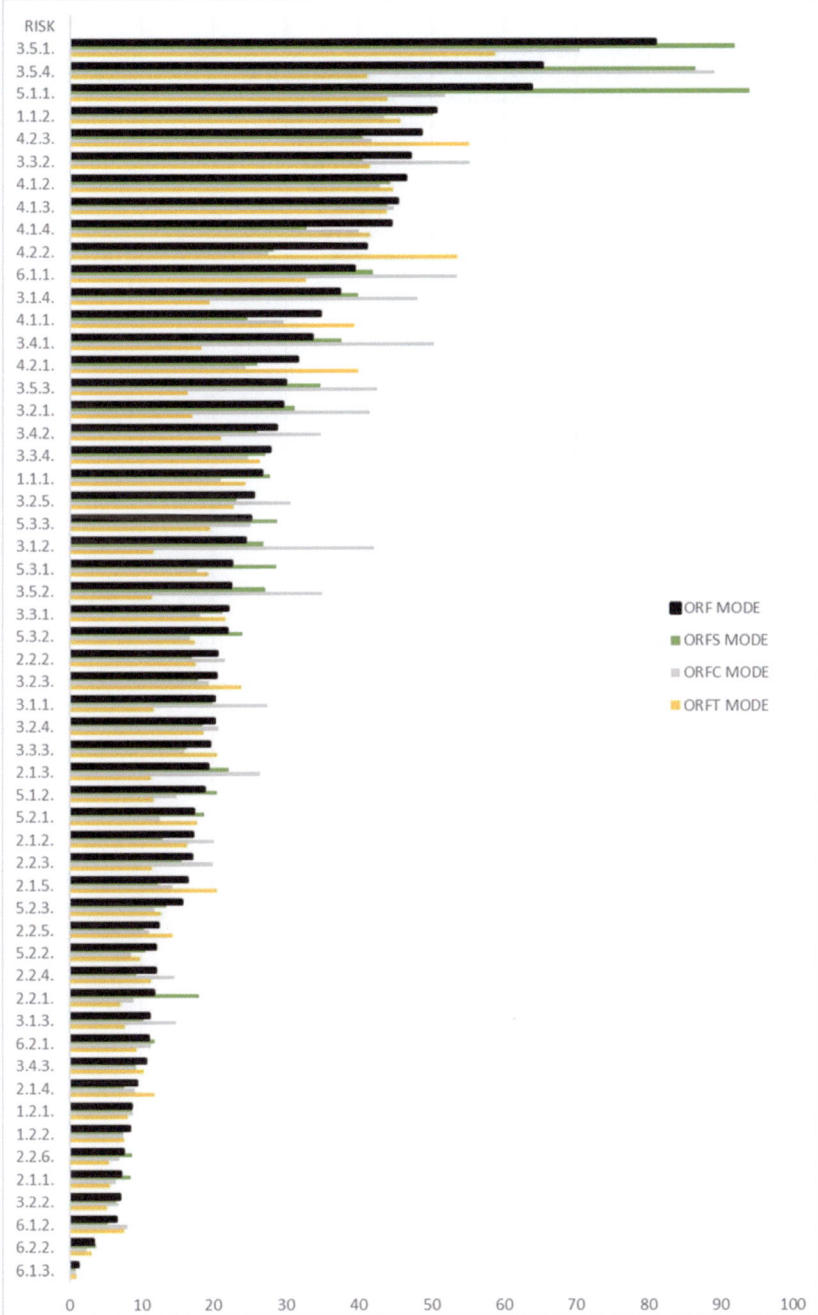

Fig. 2.7 Mode ORF risks ranking. *Source* Serrano-Gomez and Munoz-Hernandez [31]

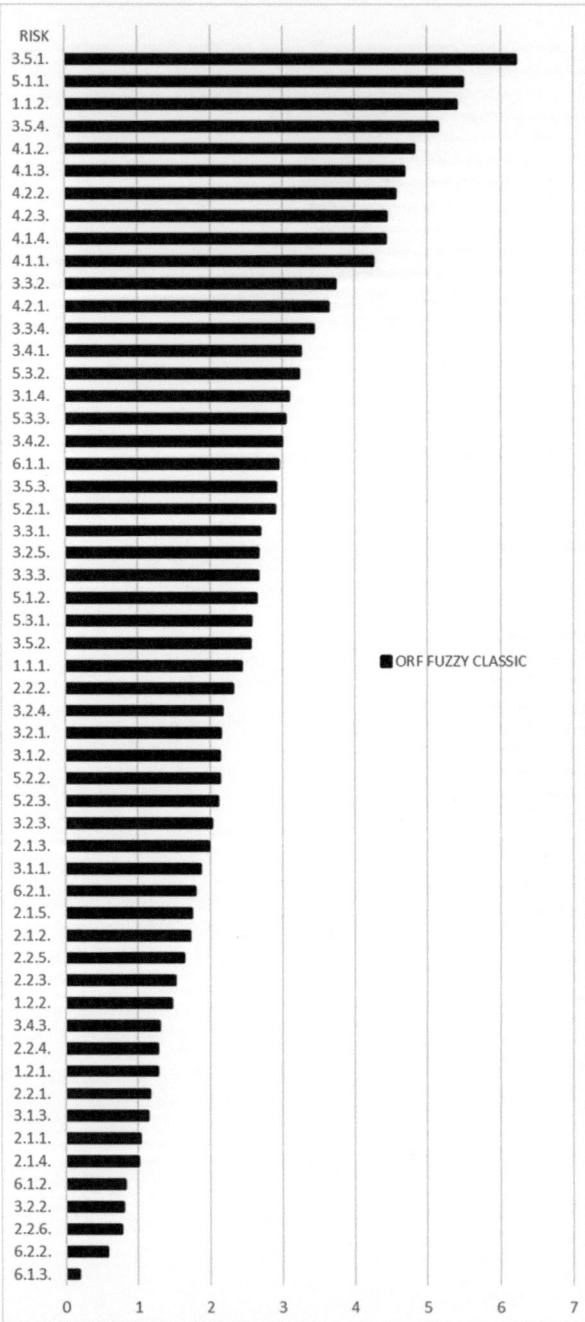

Fig. 2.8 Fuzzy classical ORF risks ranking. *Source* Serrano-Gomez and Munoz-Hernandez [31]

2.4.4 Advantages and Applications

The application of fuzzy logic based methodology is very useful in decision-making in early stage projects with a high degree of uncertainty. Some of its key advantages include:

- **Accuracy**: The combination of fuzzy logic with office tools that allow the use of distribution functions and Monte Carlo simulation such as @Risk allows for more accurate analysis than operating directly with fuzzy numbers.
- **Agility**: It is an agile and smooth application, with a fast and thorough methodology that is of great help to decision-making teams.

References

1. Mardani A, Jusoh A, Valipour A (2015) Multiple criteria decision-making techniques and their applications - a review of the literature from 2000 to 2014. Econ Res-Ekon Istraživanja 28(1):516–571. https://doi.org/10.1080/1331677X.2015.1075139
2. Aguarón J, Moreno-Jimenez J (2003) The geometric consistency index: approximated thresholds. Eur J Oper Res 147:137–145
3. Amudha M, Ramachandran M, Saravanan V, Anusuya P, Gayathri R (2021) A study on topsis mcdm techniques and its application. Data Anal Artif Intell 1(1):09–14
4. Bonissone PP, Subbu R, Lizzi J (2009) Multicriteria decision making (mcdm): a framework for research and applications. IEEE Comput Intell Mag 4(3):48–61
5. Bottani E, Rizzi A (2006) A fuzzy topsis methodology to support outsourcing of logistics services. Supply Chain Manag: Int J 11(4):294–308
6. Chen CH (2020) A novel multi-criteria decision-making model for building material supplier selection based on entropy-ahp weighted topsis. Entropy 22(2). https://doi.org/10.3390/e22020259, https://www.mdpi.com/1099-4300/22/2/259
7. Chodha V, Dubey R, Kumar R, Singh S, Kaur S (2022) Selection of industrial arc welding robot with topsis and entropy mcdm techniques. Mater Today: Proc 50:709–715. https://doi.org/10.1016/j.matpr.2021.04.487, https://www.sciencedirect.com/science/article/pii/S221478532103412X, 2nd International Conference on Functional Material, Manufacturing and Performances (ICFMMP-2021)
8. Cooper O (2017) The magic of the analytic hierarchy process (ahp). Int J Anal Hierarchy Process 9(3)
9. Diakoulaki D, Antunes CH, Gomes Martins A (2005) Mcda and energy planning. Multiple criteria decision analysis: state of the art surveys, pp 859–890
10. Dubois D, Prade H (1978) Operations on fuzzy numbers. Int J Syst Sci 9:613–626
11. Garcia-Ayllon S, Hontoria E, Munier N (2022) The contribution of mcdm to sump: the case of spanish cities during 2006–2021. Int J Environ Res Public Health 19(1). https://doi.org/10.3390/ijerph19010294, https://www.mdpi.com/1660-4601/19/1/294
12. Gil García IC, et al (2020) Integración del recurso eólico marino en los sectores del transporte y climatización: estudio de transición energética en la costa este de eeuu. Ph.D. Thesis, Universidad Politécnica de Cartagena, available online: https://repositorio.upct.es/handle/10317/9164
13. Gil-García IC, Ramos-Escudero A, García-Cascales M, Dagher H, Molina-García A (2022) Fuzzy gis-based mcdm solution for the optimal offshore wind site selection: the gulf of maine case. Renew Energy 183:130–147. https://doi.org/10.1016/j.renene.2021.10.058

14. Gil-García IC, Ramos-Escudero A, Molina-García Ángel, Fernández-Guillamón A (2023) Gis-based mcdm dual optimization approach for territorial-scale offshore wind power plants. J Clean Prod 428:139484. https://doi.org/10.1016/j.jclepro.2023.139484 https://www.sciencedirect.com/science/article/pii/S0959652623036429
15. Gil-García IC, Fernández-Guillamón A, García-Cascales MS, Molina-García A, Dagher H (2024) A green electrical matrix-based model for the energy transition: Maine, usa case example. Energy 290:130246. https://doi.org/10.1016/j.energy.2024.130246
16. Kaufmann A, Gupta MM (1991) Introduction to Fuzzy Aritmetic. Van Nostrand Reinhold Company
17. Kaya I, Çolak M, Terzi F (2018) Use of mcdm techniques for energy policy and decision-making problems: a review. Int J Energy Res 42(7):2344–2372
18. Lozano JMS (2013) Búsqueda y evaluación de emplazamientos óptimos para albergar instalaciones de energías renovables en la costa de la región de murcia: combinación de sistemas de información geográfica (sig) y soft computing. Tesis doctoral, Universidad Politécnica de Cartagena, https://dialnet.unirioja.es/servlet/tesis?codigo=51908&info=resumen&idioma=SPA, https://dialnet.unirioja.es/servlet/tesis?codigo=51908
19. Lyons T, Skitmore M (2004) Project risk management in the queensland engineering construction industry: a survey. Int J Proj Manag 22(1):51–61
20. Majumder M, Majumder M (2015) Multi criteria decision making. Impact of urbanization on water shortage in face of climatic aberrations, pp 35–47
21. Mukhametzyanov I (2021) Specific character of objective methods for determining weights of criteria in mcdm problems: Entropy, critic and sd. Decis Mak: Appl Manag Eng 4(2):76–105. https://doi.org/10.31181/dmame210402076i, https://dmame-journal.org/index.php/dmame/article/view/194
22. Munier N (2021) Linear Programming and the SIMUS Method. Springer, Cham, pp 7–14. https://doi.org/10.1007/978-3-030-82347-4_2, https://doi.org/10.1007/978-3-030-82347-4_2
23. Munier N, Hontoria E et al (2021) Uses and Limitations of the AHP Method. Springer
24. Nieto-Morote A, Ruz-Vila F (2011) A fuzzy approach to construction project risk assessment. Int J Proj Manag 29:220–231
25. Opricovic S, Tzeng GH (2004) Compromise solution by mcdm methods: a comparative analysis of vikor and topsis. Eur J Oper Res 156(2):445–455. https://doi.org/10.1016/S0377-2217(03)00020-1, https://www.sciencedirect.com/science/article/pii/S0377221703000201
26. Papathanasiou J, Ploskas N, Papathanasiou J, Ploskas N (2018) Topsis. Springer
27. Prasad RD, Bansal R, Raturi A (2014) Multi-faceted energy planning: a review. Renew Sustain Energy Rev 38:686–699
28. Ramos-Escudero A, Magraner T, Gil-García IC (2024) Optimized spatial tool for the implementation of ground source heat pump coupled with photovoltaic panels heating systems in urban areas. Energy Build 323:114752. https://doi.org/10.1016/j.enbuild.2024.114752, https://www.sciencedirect.com/science/article/pii/S0378778824008685
29. Roszkowska E, Filipowicz-Chomko M (2024) A multi-criteria method integrating distances to ideal and anti-ideal points. Symmetry 16(8):1025
30. Saaty TL (1980) The analytic hierarchy process (ahp). J Oper Res Soc 41(11):1073–1076
31. Serrano-Gomez L, Munoz-Hernandez J (2019) Monte Carlo approach to fuzzy ahp risk analysis in renewable energy construction projects. PLoS ONE 14(6):20
32. Shannon CE (1948) A mathematical theory of communication. Bell Syst Tech J 27(3):379–423. https://doi.org/10.1002/j.1538-7305.1948.tb01338.x
33. Shih HS, Olson DL (2022) TOPSIS and its extensions: a distance-based MCDM approach, vol 447. Springer
34. Stoilova S, Munier N (2021) A novel fuzzy simus multicriteria decision-making method. an application in railway passenger transport planning. Symmetry 13(3). https://doi.org/10.3390/sym13030483, https://www.mdpi.com/2073-8994/13/3/483
35. Thakkar JJ (2021) Multi-criteria decision making, vol 336. Springer

References

36. Wang CN, Le TQ, Chang KH, Dang TT (2022) Measuring road transport sustainability using mcdm-based entropy objective weighting method. Symmetry 14(5). https://doi.org/10.3390/sym14051033, https://www.mdpi.com/2073-8994/14/5/1033
37. Xu Z, Zhang X (2013) Hesitant fuzzy multi-attribute decision making based on topsis with incomplete weight information. Knowl-Based Syst 52:53–64
38. Zadeh LA (1965) Fuzzy sets. Inf Control 8:338–353

Chapter 3
Geographical Information Systems

Abstract This chapter highlights the importance of Geographic Information Systems (GIS) in assessing renewable energy sources (RES) potential and supporting informed decision-making. GIS enables stakeholders to analyze resource availability, environmental conditions, and economic feasibility through spatial data and mapping. By estimating the RES potential, identifying deployment constraints, and quantifying CO_2 reductions, GIS aids in strategic planning for RES projects. The chapter explores key GIS applications, including the IRENA Global Atlas and NASA's POWER Project, which provide RES maps at global and regional scales. GIS combined with Multi-Criteria Decision-Making (MCDM) is shown to be particularly effective for selecting locations for RES technologies, such as wind power plants and hybrid systems, with suitability maps guiding deployment. GIS-based methods help overcome these by assessing resource potential and supporting policies to raise awareness. Through case studies at continental, regional, and urban scales, GIS's role in identifying high-potential sites and optimizing SGE deployment is illustrated, underscoring its value in achieving sustainable energy goals.

3.1 GIS-Based Spatial Analysis for RES

Using Geographical Information Systems (GIS) has become crucial for evaluating the potential of Renewable Energy Sources (RES) investments. GIS allows vast amounts of data to be gathered, managed, and analysed, providing both statistical insight and graphical representation through maps. These maps highlight resource availability, environmental conditions, and the economic feasibility of deploying renewable energy technologies in specific areas. Examples include the National Renewable Energy Laboratory (NREL), who developed the RES potential maps for the United Stated (U.S.), showing state-level energy potential across wind, solar, geothermal, and other resources [33], and the International Renewable Energy Agency (IRENA), who developed the IRENA's Global Atlas for RES, which centralizes global data on wind, solar, geothermal, biomass, and hydroelectric energy resources [26].

GIS helps stakeholders like governments, utilities, and private investors make informed decisions by providing detailed spatial analysis, essential for integrating RES into policy and development strategies. However, to effectively integrate RES into various scenarios, it is essential to assess existing opportunities and plan accordingly. GIS-based spatial methodologies typically follow these key steps:

1. Estimation of renewable energy potential
2. Evaluation of local characteristics that influence this potential
3. Identification of constraints that limit the deployment of specific technologies
4. Projection of economic benefits
5. Quantification of CO_2 emission reductions at the local level.

In the field of RES, GIS plays a crucial role in managing complex data to assess a region's energy potential, identify optimal sites, allocate energy resources, and evaluate infrastructure needs. GIS supports not only the technological and economic aspects, but also the integration of RES within environmental and social contexts. This involves assessing the interactions between these domains and the technological development, physical implementation, and social adoption of RES technologies. Furthermore, GIS enables an in-depth analysis of the initial conditions, capturing and organizing relevant socioeconomic, environmental, and energy-related data. This includes information on the physical and biological environment, energy resources, population distribution, infrastructure, legal frameworks, and regional development plans. GIS has been extensively utilized in RES technologies and related disciplines across the globe. Many of these applications are accessible through online platforms featuring interactive map viewers. Table 3.1 presents a collection of the most comprehensive and widely regarded global RES map viewers.

Additionally, numerous studies have sought to assess RES potential and technologies worldwide for various purposes. For instance, Ramachandra and Shruthi [39] mapped the potential of major RES in India, providing a foundation for integrating renewable energies into regional budgets by identifying and quantifying the spatial and climatic factors influencing RE potential. In the context of wind energy, Noorollahi et al. [34] analysed the wind potential in eastern Iran, creating a wind power plant suitability map at a regional scale that indicated 28% of the area as favourable for large wind power plant installations. Integrating GIS-based methodologies with Multi-Criteria Decision-Making (MCDM) approaches is also a common practice. For example, Sáanchez-Lozano et al. [43, 44] applied a combination of GIS and MCDM techniques to select solar power plants and onshore wind power plant locations (respectively) in southeast Spain. Hybrid technologies have also benefited from these practices. Aydin et al. [3] introduced a novel GIS-based methodology for selecting sites for hybrid wind and PV systems in Turkey. Their results were represented through location prioritization maps based on the RES potential. Similarly, Messaoudi et al. [31] employed a MCDM tool combined with GIS to select sites for green hydrogen production, resulting in a suitability map with four categories ranging from low to high. Moreover, shallow geothermal energy (SGE) is beginning to transition out of its niche market, a change that has been largely unnoticed in

3.1 GIS-Based Spatial Analysis for RES

Table 3.1 Notable global RES online map viewers

Source	Viewer maps description	Resources covered	Scale
IRENA—Global Atlas for Renewable Energy [27]	RES potential maps covering solar, wind, geothermal, and bioenergy at global, regional, and national levels	SOL[a], W[b], OCE[c], HYD[d], GEO[e], BIO[f]	Global
NASA—The POWER Project [32]	Solar and wind potential maps providing surface meteorology and solar energy estimates from satellite observations	SOL, W	Global
Joint Research Centre—PVGIS [28]	Photovoltaic (PV) potential and solar radiation maps for Europe, Africa, and Asia	SOL, PV	Continental (Europe, Africa, Asia)
NREL—RES Atlas [36]	U.S.-focused RES atlas providing data on deep geothermal, solar, wind, and bioenergy	GEO, SOL, W, BIO	U.S. National
Marine Cadastre [30]	Thematic maps for offshore wind potential in U.S. coastal areas	Offshore W	U.S. National
GeoEnergy Europe—Geo-Energy Database [15]	A comprehensive database that shows geothermal potential on a global scale, with a focus on Europe. The map includes various geothermal resources and technologies, from direct use to electricity production	GEO	Global/European

[a] Solar
[b] Wind
[c] Ocean
[d] Hydrothermal
[e] Geothermal
[f] Biomass

the energy sector over the past few decades. Like many emerging technologies, SGE requires time and investment to enhance its cost-efficiency and to address the barriers that hinder its broader deployment. Several studies have identified key challenges for SGE [5, 13, 16, 20, 49], which can be classified into technical and non-technical categories. On the technical front, significant challenges remain, particularly regarding the installation of SGE systems in existing urban structures. These issues often stem from spatial limitations and constraints related to underground space, as well as operational challenges and concerns regarding resource extraction. In contrast, non-technical barriers critically impede necessary investment in SGE, further hindering its deployment.

3.2 Geo-information for Building Institutional Capacity in Renewable Energy

RES play a vital role in the EU's energy transition, and their utilization needs to triple to meet the EU's renewable energy targets for 2030. However, the progress of various RES technologies can be hindered by technical and institutional preferences for more established options, as well as public resistance. Similar challenges have been observed with wind and solar energy, where certain segments of society view these technologies as culturally and environmentally intrusive. Fortunately, these barriers can be addressed when linked to a common objective: enhancing policy capacity to achieve social acceptability and addressing economic and environmental constraints [6]. The effectiveness of government initiatives to foster a renewable future depends significantly on the quality of baseline data regarding resource availability [7]. At the political level, there remains a lack of understanding regarding the sustainable and economic potential of various RES to deliver energy services, as well as the capacity of new systems to meet established policy goals related to land use, environmental management, socio-economic development, and energy security. Moreover, there is no single "miraculous" RES technology; thus, governments must consider and integrate a diverse array of RES into their policies and energy strategies [12].

Most RES and their conversion systems are highly influenced by geographical constraints and the regional political and economic context [9, 46]. The integration of GIS with energy management is widely acknowledged in academic circles as essential for developing this baseline information [2, 11, 47]. Key values of geographic information include:

- Providing numerical data to assess existing resource stocks in specific areas, adaptable to various geographical scales [11].
- Offering cartographic representations that facilitate comprehension and serve as effective communication tools for policymakers and managers [11].
- Creating numerous opportunities for direct collaboration through web mapping, which can enhance public engagement in energy planning [25].

This section emphasizes the application of spatial information tools in decision support and demonstrates how effective information management can address the scarcity of data surrounding RES. The results can be integrated into energy and planning strategies. Figure 3.1 illustrates the communication links among key stakeholders involved in RES in the EU and the crucial role of GIS in this process.

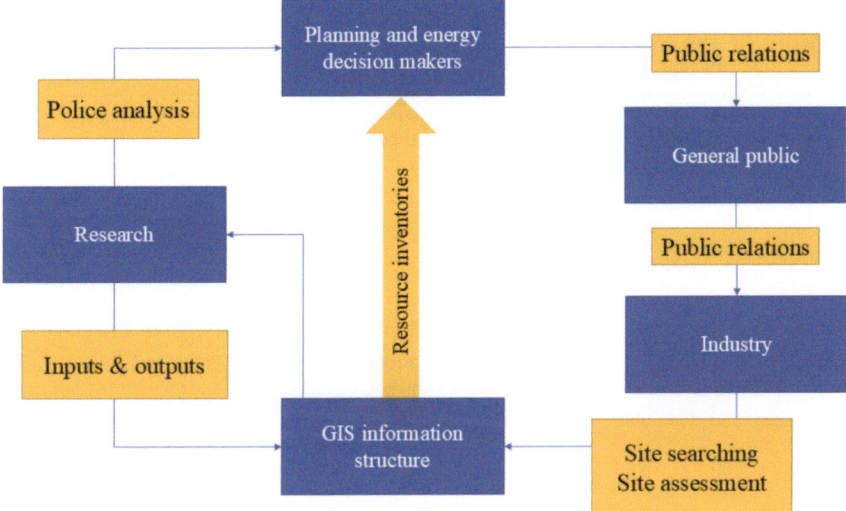

Fig. 3.1 Communication links between stakeholders and GIS information to feed into energy policies. *Source* Calvert et al. [7]

3.3 GIS-Based Decision Support

Technically, spatial analysis in the field of RES follows a similar pattern across various technologies. However, some RES often face significant challenges due to a lack of comprehensive information and insufficient data dissemination. This situation contributes to low public awareness and limited political support, as illustrated in Table 3.2. To address these barriers, GIS-based spatial analysis plays a crucial role in evaluating RES potential. This analysis not only highlights how much-untapped energy can be harnessed from various sources, but also clarifies the economic and environmental conditions necessary for their viability. Over the past few decades, extensive spatial analysis efforts have been undertaken with a shared objective: to promote the adoption of RES.

The resulting maps serve a dual purpose: (i) they can either provide immediate solutions to decision-makers, or (ii) function as decision-support tools, such as thematic maps. Thematic maps are particularly valuable, as they consolidate relevant information, enhance understanding, and provide insights that aid the decision-making process. By visually representing complex data, these maps can effectively communicate the benefits and feasibility of RES, thereby fostering greater awareness and support for their implementation.

In this section, potential maps, conflict use maps, optimal location selection, and fuzzy analysis are explored to understand how they contribute to a clearer understanding and better management of RES. Through these elements, it is possible to

Table 3.2 Main non-technological barriers of RES. *Source* GEOPLASMA Project [19] and GEO4CIVHIC [18] Project

	Barrier	Proposed solutions
1	Complexity on legal framework	Simplification and unification of legal regulations
		Supporting administrative procedures and licensing
		Harmonized national funding
2	High upfront costs	Shallow geothermal use connected
		Reduce CAPEX
3	Very low public awareness	Provide strong support for specialist groups
		Promote the shallow geothermal acknowledgment
		Spread the knowledge. Information campaigns
		Include SGE in energy strategies and planning tools
4	Limited access to the information	Harmonize technical terms and methodologies
		Provide web base access information
5	Limited qualifies access	Support standardized qualifications schemes
6	Unknown detailed market information	Create regional and national register

identify opportunities, mitigate risks, and facilitate sustainable development in the use of the different RES. Geothermal energy is used as an example along the section.

When evaluating any RES potential (geothermal), it is essential to distinguish between resource, technical, economic, and market potential, as illustrated in Fig. 3.2.

3.3.1 Resource Potential Maps

They serve as an initial estimate of the energy available for extraction through current technology, calculated over time at a specific location and depth. This is a fundamental step in energy mapping, where energy units like GW h/year are typically used. However, in the specific case of RES, these maps aren't suitable for designers, since they lack other critical factors. In the specific case of geothermal systems, resource potential maps depend on localized underground characteristics, and the most frequently sought-after data types include thermal conductivity and heat extraction maps. Secondary, though still valuable, are maps detailing hydrogeological factors, heat flux, and water table depth.

3.3 GIS-Based Decision Support

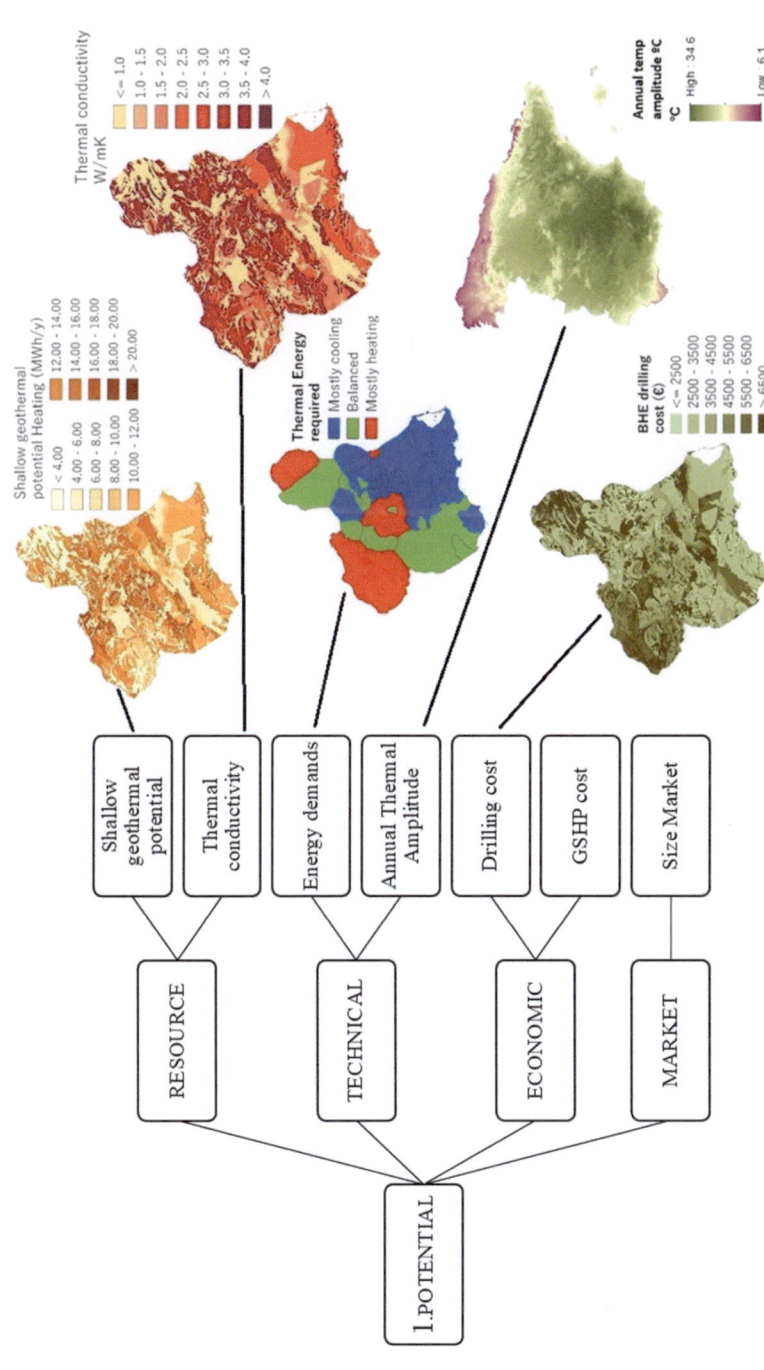

Fig. 3.2 Potential map classification and graphic examples. Source from top to bottom: Ramos-Escudero and Socorro García-Cascales [40] and Ramos-Escudero et al. [41]

3.3.2 Technical Potential Maps

They measure the achievable energy production capacity in given locations or regions based on technology capabilities, performance standards, topographic constraints, and environmental or land-use limitations. For instance, for geothermal energy, technical evaluations often consider factors like site-specific energy demands for heating and cooling or the Coefficient of Performance (COP) of shallow geothermal energy systems.

3.3.3 Economic Potential Maps

They focus on the cost-efficiency of harnessing shallow geothermal energy in particular locations. Significant parameters in evaluating geothermal economics include drilling expenses and the costs associated with replacing commonly used technologies.

3.3.4 Market Potential Maps

They assess the economic feasibility of adopting RES technology in specific regions, projecting market share estimates and economic benefits. This assessment incorporates resource, technical, and economic potential data, along with population and environmental advantages. For geothermal energy, Hellström and Sanner [24] provided early models for estimating the heat extraction capacity of boreholes, but Ondreka et al. [37] pioneered the first GIS-based shallow geothermal map of southwestern Germany for a closed-loop heating system. This work has inspired further studies across various global regions, such as those by Carranza et al. [8], Blum et al. [4], Di Sipio et al. [10], Galgaro et al. [16], Luo et al. [29].

Although most GIS-based studies are resource-focused, some extent into broader potential mapping. For example, Gemelli et al. [17] developed a regional-scale model incorporating energy economics, addressing factors like installation cost, payback time, and geothermal energy pricing. Blum et al. [4] quantified regional CO_2 reductions from switching to geothermal heating. Noorollahi et al. [35] evaluated Iran's geothermal potential across the resource, technical, and economic dimensions, producing a priority map for ground source heat pump (GSHP) systems to guide public investment decisions. More recently, Walch et al. [50] examined thermal effects associated with large-scale borehole heat exchanger installations, assessing technical potential across Switzerland.

3.3 GIS-Based Decision Support

3.3.5 Conflict Use Maps

Conflict use maps highlight the limitations and constraints that can make any RES unfeasible. In Fig. 3.3, the conflict use maps for shallow geothermal installations can be seen. These maps restrict shallow resource accessibility by representing activities that may cause environmental harm or interfere with nearby land uses, directly or indirectly. For example, Hamada et al. [23] analysed open- and closed-loop system conditions in Japan. After examining hydrogeological data and environmental regulations, they found that only 33% of the land could be considered suitable for these systems. Standard conflict use maps also support the creation of suitability maps and "traffic light" maps.

Suitability maps indicate the suitability level of specific technologies within an area, while traffic light maps indicate whether installation is viable, restricted, or prohibited. These maps integrate energy-related factors with RES parameters and local constraints, harmonized for comparison purposes. A colour or label (such as low, medium, or high suitability, or red, orange, and green) is assigned based on the technology's projected success in a given location, offering valuable insights for regional and local energy decision-makers. Regarding geothermal, the EU-funded GEOPLASMA project (2016–2019) evaluated geothermal resources and potential conflicts across six pilot areas in Central Europe. The project produced a web-GIS tool where conflict use maps could be reviewed, accounting for areas under nature protection, water protection, contamination risk, and groundwater zones suited for open-loop systems. Separate suitability maps were created for groundwater heat pumps and borehole heat exchangers.

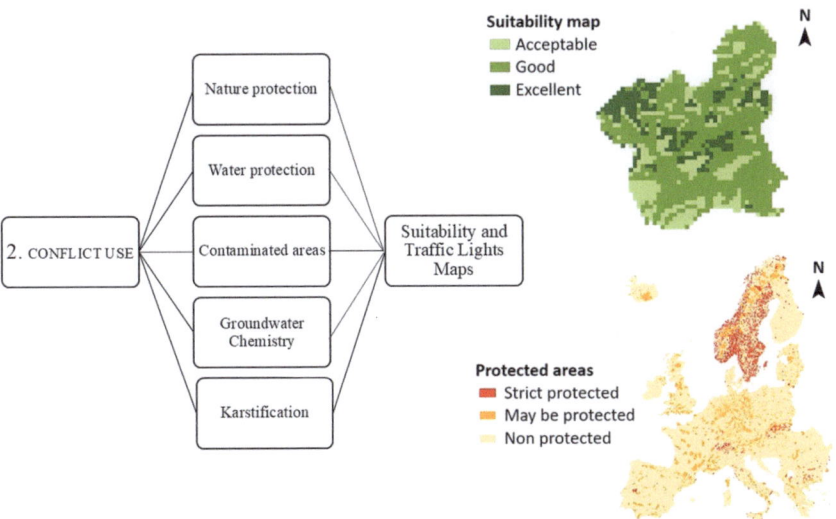

Fig. 3.3 Conflict uses maps classification and examples. Source of the maps: Ramos–Escudero et al. [41]

3.4 GIS-Based Mapping Typical Workflow

In GIS-based mapping studies, a standard workflow is generally followed for data handling and spatial information processing, regardless of the specific methodology. Figure 3.4 outlines the main steps involved in creating decision-support maps for RES.

The process begins by defining the map's objective, which determines its type among those previously described. The next phase involves data acquisition, with Steps 1 to 4 (Fig. 3.4, Data section) focusing on gathering primary data essential for the mapping objective. Once the map's purpose is established, the available data defines the map's technical characteristics, such as resolution and scale, and may even influence the map's feasibility. High-resolution primary spatial data acquisition and accurate modelling are key for ensuring precision and reliability in the final geophysical process models [21]. High-resolution data is always preferred, being two main types of spatial data models typically used: vector and raster. Often, data in those two formats must be converted to a common model to ensure compatibility, with raster models frequently used in analysis involving Map Algebra—a widely used GIS tool—[45]. In cases where multiple data layers overlap, the final map resolution defaults to that of the lowest resolution layer. Any data gaps are also noted; for maps at national or continental scales, specific regional data may be missing, which can prevent map completion. Once data availability is confirmed, data collection proceeds with the chosen GIS tool.

The second phase involves all GIS-specific operations and data manipulations needed to generate the map. A variety of GIS software options exists, including

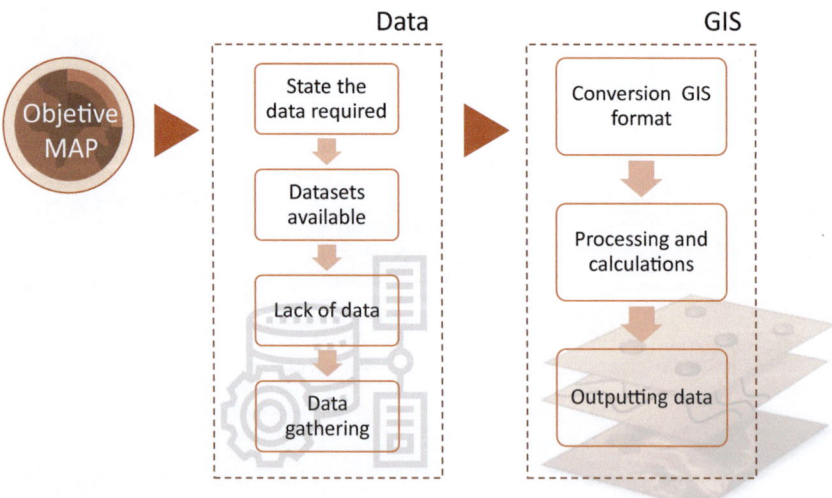

Fig. 3.4 Typical workflow to produce decision-making maps

Table 3.3 Most used tools in a GIS software for RES mapping

Model format	Tool package	Tools
Vector	Geoprocessing	Buffer, Dissolve, intersection, merges
	Geometry	Line intersection, nearest neighbour's analysis
	Analysis tools	Create grid, random points ain extent
	Data management	Join tools, vector layers merge
Raster	Raster calculator	Map algebra
	Proximity, slope	
	Extraction	Raster cut by extension or layer mask
	Proximity, slope	
	Conversion	Polygonize, rasterized

open-source tools like SAGA GIS [42], ILWIS [1], gvSIG [48], GRASS GIS [22], ArcGIS [14], and QGIS [38]. ArcGIS is one of the most widely used GIS tools.

Steps 5 to 7 of Fig. 3.4 outline GIS-based tasks. Step 5 may not always be required if data is already in a GIS-compatible format but, when conversion is necessary, it typically involves transforming raster-to-vector or vector-to-raster formats as needed. Step 6 includes executing GIS calculations and applying functions to achieve the desired map outcomes. The standard GIS tools for these tasks are listed in Table 3.3.

3.5 From Continental to Regional Scales: GIS-Based Studies of Renewable Energy Potential

GIS and MCDM-based mapping at all levels (continental, regional, and local) serves as a powerful tool to illuminate the RES potential in Europe, aligning technical insights with policy implementation and community awareness to drive sustainable development. At each scale, producing spatially-informed, GIS-based potential maps enables informed decision-making regarding the resource, technical, and socio-economic potential of RES.

Each scale uniquely enhances renewable energy studies: the lcontinental scale assesses broad resource potential and informs EU-wide strategies, fostering international alignment. National to regional scale narrow this focus, considering geographic and socio-economic factors to support targeted energy planning, address regional challenges, and optimize resource distribution. Finally, the local scale provides detailed, site-specific insights on feasibility, cost-effectiveness, and technology choice, translating broader policies into actionable, community-level solutions, ultimately strengthening renewable energy adoption across all levels.

3.5.1 Continental Scale

On a continental scale, there is a pressing need to provide comprehensive information on RES potential to enable their widespread adoption across the EU. Through GIS-based methodologies, thematic maps can capture variables affecting the system suitability, reflecting the influence of several conditions. These maps are vital for illustrating RES suitability across Europe. The primary goals of continental-scale RES potential mapping include:

- Offering the scientific community detailed insights into the different critical parameters for effective RES operation.
- Expanding public knowledge on the potential of RES.
- Providing decision-makers with valuable spatial information and statistical evidence highlighting the benefits of RES for European communities.
- Identifying areas within the EU with significant resource, technical, and economic potential for RES.
- Encouraging the integration of RES into EU energy policies, helping align national energy plans with sustainable development goals.

3.5.2 National to Regional Scale: MCDM-Supported Suitability Maps

Energy planning and management often require more granular insights at national and regional scales, where decision-makers face unique geographic and policy-related challenges. The decentralization trend in European energy governance increasingly entrusts local authorities with decision-making on RES deployment. Consequently, creating GIS-based suitability maps at this scale can support regional energy managers by providing a robust framework for assessing RES potential. The objectives of regional-scale mapping include:

- Supplying regional decision-makers with a comprehensive energy-environment-economic knowledge base to guide RES adoption.
- Increasing awareness among regional authorities and the public about the presence of RES, their cost-effectiveness, and environmental benefits as tools against climate change.
- Contributing innovative insights to the scientific community on RES potential in region-specific conditions.
- Identifying the primary barriers hindering the effective deployment of RES in specific regions.

By integrating MCDM with GIS-based mapping, regional maps can prioritize areas for RES based on criteria weighted for their impact on the system performance.

3.5.3 Local or Urban Scale

At the local or urban scale, the decentralized nature of RES applications makes it essential to translate EU policy into localized actions. Here, local action plans foster economic growth by implementing EU directives, with GIS-based analyses helping decision-makers understand site-specific characteristics. The GIS approach at this scale is aimed at:

- Raising public and decision-maker awareness of the local RES resource.
- Evaluating the energy production and greenhouse gas reduction benefits associated with RES deployment.
- Assessing the cost-effectiveness of RES systems based on local conditions.
- Identifying and addressing barriers to more extensive RES deployment.
- Recommending the optimal RES technology tailored to the local context.

References

1. 52north (2007) ILWIS: integrated land and water information system. https://www.ilwis.org, accessed: 2023-10-29
2. Angelis-Dimakis A et al (2011) Methods and tools to evaluate the availability of renewable energy sources. Renew Sustain Energy Rev 15(2):1182–1200
3. Aydin NY, Kentel E, Duzgun HS (2013) Gis-based site selection methodology for hybrid renewable energy systems: a case study from turkey. Energy Convers Manag 70:90–106
4. Blum P et al (2010) Quantifying co2 savings of geothermal heating systems in germany. Appl Energy 87:3211–3216
5. Breembroek G, Ramsak P, Manzanella A (2014) Challenges and barriers in shallow geothermal energy development: Technical and non-technical aspects. Geotherm Energy 2(1):1–15
6. Burkard E, Doern GB (2010) Social acceptability and policy capacity in the energy sector. University of Toronto Press
7. Calvert K, Pearce JM, Mabee WE (2013) Renewable energy and the policy landscape in north america: opportunities and barriers. Renew Sustain Energy Rev 23:149–160
8. Carranza EJM et al (2008) Geothermal resource assessment using gis techniques in the philippines. J Volcanol Geotherm Res 175:675–689
9. Cassedy ES (2000) Prospects for sustainable energy: a critical assessment. Cambridge University Press
10. Di Sipio E et al (2014) Mapping geothermal resources in italy using gis: case studies in tuscany. Renew Sustain Energy Rev 30:180–195
11. Domínguez J, Amador J (2007) Gis-based assessment of sustainable energy potentials in andalusia. Renew Energy 31(14):2366–2382
12. Edenhofer O, et al (2011) Special report on renewable energy sources and climate change mitigation. Intergovernmental panel on climate change (IPCC)
13. Emmi G, Zarrella A (2020) Challenges and strategies for shallow geothermal deployment. Geothermics 87:101929
14. Environmental Systems Research Institute (2011) ArcGIS Desktop: Release 10. Redlands, CA
15. EuroGeoSurveys (2024) GeoEnergy Europe. https://www.europe-geology.eu/scientific-themes/geoenergy/, https://www.europe-geology.eu/scientific-themes/geoenergy/, accessed: 2024-12-26

16. Galgaro A et al (2015) A gis-based assessment of shallow geothermal energy potential in europe. Renew Energy 83:1–13
17. Gemelli A, Mancini R, Longhi S (2011) Gis-based model for the economic assessment of geothermal energy at a regional scale. Renew Energy 36:2337–2344
18. GEO4CIVHIC (2024) Geo4civhic: most easy, efficient and low cost geothermal systems for retrofitting civil and historical buildings. https://geo4civhic.eu/, accessed: 2024-12-26
19. GEOPLASMA Project (2019) Geoplasma: Web-gis tool for assessing geothermal resources and environmental conflicts in central europe. https://www.interreg-central.eu/Content.Node/GEOPLASMA-CE.html, https://www.interreg-central.eu/Content.Node/GEOPLASMA-CE.html, eU-funded project, 2016–2019
20. Goetzl G, Heiermann I (2018) Geothermal barriers and potential for development in germany. Energy Policy 113:1–7
21. Goodchild MF (1992) Geographical information systems and science. Wiley
22. GRASS Development Team (2017) GRASS GIS: a free and open source geographic information system. https://grass.osgeo.org, accessed: 2023-10-29
23. Hamada Y, Marutani Y, Nakamura M (2003) Suitability analysis of shallow geothermal systems for open- and closed-loop systems in japan. Geothermics 32(2):239–247
24. Hellström G, Sanner B (1994) Earth energy systems with heat extraction boreholes: models and applications. Geothermics 23(4):389–404
25. Higgs G et al (2008) Web-based gis for public engagement in the renewable energy sector. Environ Plan B: Plan Des 35:493–511
26. International Renewable Energy Agency (n.d.) The international renewable energy agency (irena). https://www.irena.org, accessed: 2023-10-29
27. IRENA (2024) Global atlas for renewable energy. https://globalatlas.irena.org/workspace, https://globalatlas.irena.org/workspace, accessed: 2024-12-26
28. Joint Research Centre (2024) Photovoltaic Geographical Information System (PVGIS). https://joint-research-centre.ec.europa.eu/photovoltaic-geographical-information-system-pvgis_en, https://joint-research-centre.ec.europa.eu/photovoltaic-geographical-information-system-pvgis_en, accessed: 2024-12-26
29. Luo J et al (2018) Assessing geothermal potential in china with gis-based analysis. Renew Energy 119:245–258
30. Marine Cadastre Hub (2024) Marine Cadastre Hub. https://hub.marinecadastre.gov/, https://hub.marinecadastre.gov/, accessed: 2024-12-26
31. Messaoudi D et al (2019) Mcdm and gis-based approach for green hydrogen production site selection. Int J Hydrog Energy 44(22):11643–11655
32. NASA (2024) The POWER Project. https://power.larc.nasa.gov/, https://power.larc.nasa.gov/, accessed: 2024-12-26
33. National Renewable Energy Laboratory (n.d.) The national renewable energy laboratory (nrel). https://www.nrel.gov, accessed: 2023-10-29
34. Noorollahi Y, Yousefi H, Mohammadi M (2016) Wind energy assessment and wind farm suitability analysis in east of iran using gis and multi-criteria decision making method. Renew Energy 114:306–318
35. Noorollahi Y, Gholami Arjenaki N, Ghasempour R (2017) Geothermal potential assessment using gis and multi-criteria decision analysis in iran. Renew Energy 114:37–48
36. NREL (2024) Marine energy atlas. https://maps.nrel.gov/marine-energy-atlas/?vL=OmnidirectionalWavePowerMerged, https://maps.nrel.gov/marine-energy-atlas/?vL=OmnidirectionalWavePowerMerged, accessed: 2024-12-26
37. Ondreka J et al (2007) Gis-based evaluation of shallow geothermal energy potential in southwestern germany for closed-loop heating systems. Renew Energy 32:2186–2199
38. QGIS Development Team (2018) QGIS: a free and open source geographic information system. https://qgis.org, accessed: 2023-10-29
39. Ramachandra T, Shruthi B (2007) Spatial mapping of renewable energy potential in india using gis. Renew Sustain Energy Rev 11(7):1460–1480

References

40. Ramos-Escudero G, Socorro García-Cascales M (2021) Classification of geothermal potential maps and practical applications in spain. Renew Energy 165:1103–1115
41. Ramos-Escudero G et al (2020) Gis-based analysis for geothermal energy development and potential mapping. Appl Energy 268:114–127
42. SAGA GIS Team (2005) SAGA GIS: system for automated geoscientific analyses. https://saga-gis.sourceforge.net, accessed: 2023-10-29
43. Sánchez-Lozano JM, García-Cascales MS, Lamata MT (2016) Fuzzy multi-criteria decision-making applied to gis-based site selection for onshore wind farms. Energy 107:231–242
44. Sánchez-Lozano JM et al (2014) Gis-based multi-criteria approach for solar farms site selection using electre and gis tools in southeast spain. Renew Energy 66:478–494
45. Silva-Coira F, Paramá JR, Ladra S, López JR (2020) Efficient raster and vector data integration in gis through map algebra. Int J Geogr Inf Sci 34(8):1622–1645
46. Stoeglehner G, Niemetz N, Kettl KH (2011) Spatial dimensions of sustainable energy systems: geographical and political challenges. Energy Policy 39:852–863
47. Sárensen B, Meibom P (1999) Gis and multi-criteria decision-making in the renewable energy sector. Renew Sustain Energy Rev 3:427–441
48. Technical University of Valencia (2010) gvSIG: open source geographic information system. https://www.gvsig.com, accessed: 2023-10-29
49. Tinti F et al (2016) Geothermal heat pump systems: barriers and technical solutions for shallow geothermal development in italy. Energy Procedia 97:257–264
50. Walch A et al (2021) Thermal effects of large-scale borehole heat exchanger installations in switzerland. Geothermics 89:101915

Chapter 4
Integration of MCDM with GIS and Case Studies

Abstract The integration of Multi-Criteria Decision Making (MCDM) and Geographic Information Systems (GIS) in the clean energy field is an innovative and powerful approach that helps make informed and efficient decisions in the planning and development of renewable energy sources (RES) projects. Both tools combine their strengths to address the complexity of choosing and prioritizing locations, technologies, and strategies based on multiple criteria, such as sustainability, costs, environmental impact, and technical feasibility. The methodology for selecting optimal sites by merging MCDM and GIS consists of 3 phases. In Phase 1, constraint and selection factors are established, data is collected and the potential area is delimited. Phase 2 includes the design of a prototype plant, identification of alternatives and the construction of the decision matrix. In Phase 3, weights are assigned to the factors and a ranking of alternatives is obtained.

4.1 Introduction

The integration of MCDM and GIS is particularly relevant in the RES field because many of the problems faced in energy projects' planning have a spatial dimension [2]. Some ways in which these two tools work together include:

- **Site selection for RES projects**: One of the most common uses of the combination of MCDM and GIS is the selection of optimal sites for the installation of RES infrastructure such as wind power plants [7], solar power plants [20], hydroelectric power plants [3], geothermal power plants [15] or biomass power plants [11]. In these cases, GIS helps to map and analyse geospatial data such as solar irradiation or wind speed, while MCDM allows to evaluate and weight the different criteria that affect the decision-making, such as costs, access to roads, or social impact.
- **Scenario modelling**: In RES projects, it is essential to model different scenarios to see the impact of various alternatives. For example, a wind power plant project may have several candidate sites. Using GIS, it is possible to visualize and analyse these sites on an interactive map and, with MCDM, these sites can be compared based on critical factors such as wind speed, proximity to power grids or land cost [8].

- **Environmental and social impact assessment**: Decisions on clean energy projects should not only consider technical or financial criteria, but also aspects related to environmental and social impact. GIS allows to overlay data layers of, among others, protected areas, population density, ecological corridors [12]. With the help of MCDM, it is possible to integrate these factors into the decision process and ensure that the sustainable solution chose minimizes the negative impacts [16].

Some typical applications of MCDM-GIS integration in RES include:

- **Solar energy**: Determining the best sites for large-scale solar panels, considering factors such as solar irradiance, terrain slope, distance to transmission lines, land use, and environmental impact.
- **Wind energy**: Evaluating and selecting locations for wind power plants based on wind speed and direction, proximity to transmission infrastructure, terrain, biodiversity, and social or political constraints.
- **Hydroelectric energy**: Identifying optimal sites for dams or hydroelectric power plants based on geographic data on river flow, topography, proximity to urban areas, and the presence of sensitive ecosystems.
- **Geothermal energy**: Select ideal locations for geothermal plants considering the thermal gradient of the subsoil, proximity to groundwater sources (for wet geothermal systems), tectonic or volcanic activity in the region, availability of infrastructure, and environmental impact.
- **Biomass**: Map and assess the availability of biomass in a given area and the infrastructure needed for its collection, transportation, and processing.

Overall, the benefits of integrating MCDM and GIS in the RES field are remarkable. This combination enables data-driven decision-making, as it facilitates decisions based on quantitative and qualitative analysis that integrate various sources of information [14]. Furthermore, it contributes to resource optimization, allowing the most efficient locations and strategies to be selected, maximizing energy performance while minimizing costs [19]. It also promotes sustainability, helping to balance economic benefits with social and environmental costs. Finally, it fosters transparency and participation, as GIS tools allow for clear and understandable visualizations, facilitating communication with both society and decision-makers [22].

Despite its benefits, the integration of MCDM and GIS in the clean energy space faces certain challenges. One of them is the availability and quality of geospatial data [9], which may not always be complete or up-to-date. There is also an inherent technical complexity, as the combination of multiple criteria and tools requires specialized knowledge for proper implementation [10]. Furthermore, uncertainty in the weights of the criteria is another challenge, as these weights are often subjective and can vary depending on the perspective of the decision-makers [13].

> The integration of MCDM and GIS in RES projects offers a robust solution to address the complexity in energy project planning. It helps balance technical, economic, environmental and social considerations, optimizing decision-making in a crucial sector for the transition to a sustainable energy future.

4.2 Methodology

Figure 4.1 presents an overview of the approach used to select optimal sites by integrating MCDM and GIS. This process is divided into three phases:

- **Phase 1**: Define the restrictive and selection factors, collect the necessary data, and delimit the potential area.
- **Phase 2**: Design a prototype plant model, identify the possible alternatives, and build the decision matrix.
- **Phase 3**: Assign weights to the selection factors and generate a ranking of the available alternatives.

4.2.1 Phase 1

Based on the study area, restrictive, and selection factors are identified. Restrictive factors limit the possibility of establishing RES facilities due to restrictions such as protected environmental zones, incompatible use with other economic activities, or protection policies. Selection factors influence decision-making to identify the best alternatives, defining ranges that are admitted or not and an ideal function (maximize or minimize) according to optimization objectives [5]. These factors, also known as criteria [17] or evaluation and exclusion criteria [21] in many references, are grouped into categories such as climatic, geographic, environmental, social, economic, political, and location [6]. Many coincide between different types of RES facilities, although some are specific, such as wind speed and direction for wind power plants, or solar irradiance for photovoltaic power plants.

For each factor, data must be collected, either in spatial format directly or in other formats (.csv, .xml, .dxf, etc.) that are then transformed into GIS layers. This process may involve additional tools, such as R (a free software environment and programming language designed specifically for statistical analysis and data manipulation) or database queries, but it always seeks the same objective: to generate layers that can be managed in GIS. In addition, new layers can be incorporated once the alternatives have been generated, as will be seen in phase 2 (see Sect. 4.2.2).

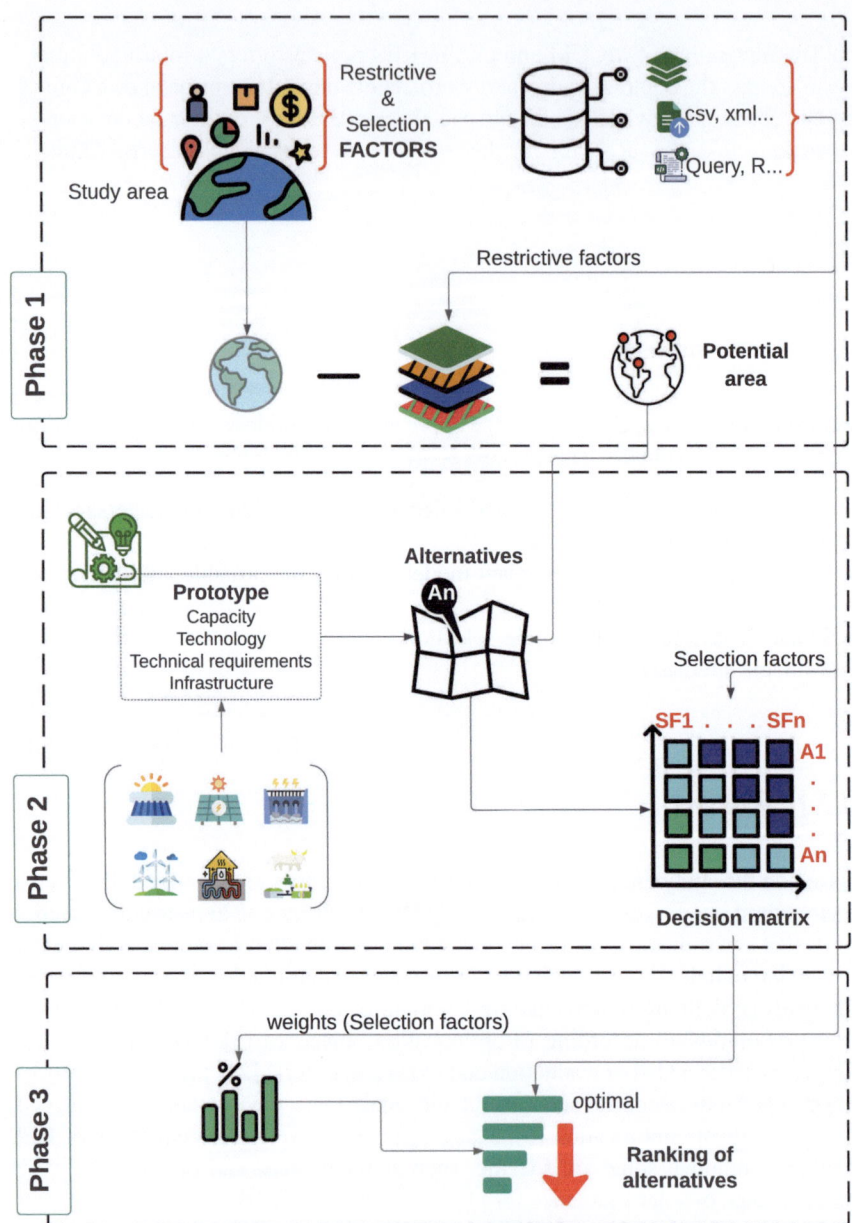

Fig. 4.1 GIS + MCDM optimal alternative selection methodology

Once all the layers have been created, the study area is reduced to a potential zone, excluding restrictive factors, such as bio-protected zones, economic activities (fishing, aquaculture), military zones, maritime routes, protected cables and pipelines, and volcanic areas.

4.2.2 Phase 2

Then, a prototype plant is designed, the configuration of which will depend on various key factors such as the projected capacity, the technology to be implemented, and the specific technical requirements. If various RES technologies are to be analysed, it will be necessary to establish multiple comparative scenarios to evaluate their implications. In addition to these aspects, other fundamental elements must be considered, such as the available infrastructure, as well as a set of indicators that will serve as the basis for the prototype design. These indicators will include both technical and economic variables, which will guide the selection process of the best alternative. It is important to note that the objective may consist of identifying alternatives or areas with different levels of suitability, i.e., to draw up a suitability map for a specific RES. In these cases, it is possible to dispense with a specific technical prototype, since the alternatives would be based on the selection criteria themselves.

This design proposal will be replicated throughout the potential area, so that for each alternative, it is possible to accurately identify how the different selection factors interact. This can be achieved both by using the existing selection factors and by computing new data, which can be integrated as new layers of information in the analysis. For example, if electricity generation is desired to be one of the selection factors, it is advisable to calculate this indicator specifically for each alternative, rather than making a general estimate for the entire potential area. The individual analysis of each alternative usually requires the use of specialized software in the sector, both for technical and economic factors, such as system losses or investment costs, among others.

At the end of the process, a decision matrix must be built, which reflects the intersection between each of the evaluated alternatives and the different selection factors previously defined. This matrix allows visualizing how the alternatives behave against the objective function of each selection factor. For example, in many cases, the goal will be to maximize production, minimize costs, and reduce losses. The decision matrix will serve as a key tool to identify the most efficient alternative; this process is analysed in the next phase.

4.2.3 Phase 3

In this phase, the weights of the selection factors are determined, with the objective of establishing the relative importance of each criterion within the evaluation process. Assigning these weights allows certain criteria to influence the final decision more than others, depending on their relevance in the specific context of the problem. In this way, the weights reflect the priorities or preferences of the decision maker or the interested groups, recognizing that not all factors have the same level of importance, see Chap. 2.

With the weights determined and the decision matrix, it is possible to apply a method to rank alternatives, see Chap. 2. This ranking facilitates the decision-making process by allowing a systematic and objective comparison of the various alternatives according to their performance against multiple factors. The main objectives of ranking the alternatives in MCDM are detailed below:

- **Identify the best alternative**: The main objective is to determine which of the available alternatives best satisfies the defined criteria. By ranking alternatives, the optimal or most efficient option can be seen, and also which are the least preferable.
- **Facilitate systematic comparison**: Ranking alternatives allows direct comparisons between them, so that it is possible to evaluate how each option performs against different criteria. This is especially useful when the alternatives have advantages and disadvantages in different aspects.
- **Prioritize decisions**: Ranking helps decision-makers prioritize the options that perform better on key factors, which can be crucial in situations where limited resources must be allocated, or strategic decisions must be made.
- **Manage complexity**: In problems where multiple criteria are evaluated (economic, environmental, technical, social, etc.), having a ranking simplifies the complexity of the analysis by summarizing the overall performance of each alternative. This helps clarify which is the most convenient option without losing sight of the balance between the evaluated criteria.
- **Explore trade-offs between criteria**: By looking at how alternatives are ranked, decision-makers can identify trade-offs. For example, one alternative might be very good in terms of costs, but not so good in terms of environmental impact, which can influence the final decision depending on the weights assigned to each criterion.
- **Justify the final decision**: Having a clear ranking of the alternatives provides an objective and transparent basis for justifying the final decision. This is especially important when the decision involves multiple stakeholders, or when quantitative and qualitative support is needed for the choice.
- **Improve decision-making under uncertainty**: In situations where there is uncertainty or lack of certainty about some factors, ranking the alternatives makes it possible to visualize which of them has a robust performance in different scenarios, facilitating the selection of the most resilient or adaptable option.

4.3 Application of GIS + MCDM

In the following sections, different applications of GIS + MCDM related to RES planning are summarized.

4.3.1 Offshore Wind Energy

4.3.1.1 Case Study: Gulf of Maine, U.S.

Gil-García et al. [7] present a MCDM approach to assess optimal offshore wind locations using fuzzy GIS. It incorporates climatic, geographic, social, environmental, and economic factors. The framework includes an AHP and a fuzzy GIS for prioritizing alternatives, with the Gulf of Maine as a case study. A 1 GW wind power plant design with 15 MW turbines is evaluated, based on 10 years of wind data. Results highlight the prioritization of optimal sites, levelized cost of electricity, and emissions reduction compared to fossil fuels.

> **Study area, restrictive and selection factors, potential area**
> Due to the lack of official maritime spatial planning in that year, the authors began by assessing the wind potential of the study area. Based on 10 years of data, they determined a high potential for offshore wind energy, with an average wind speed of 10.2 m s^{-1} at 150 m height. Using the layers of the international maritime boundary regulatory framework and the exclusive economic zone, they delimited a study area of 30,642 km^2.
>
>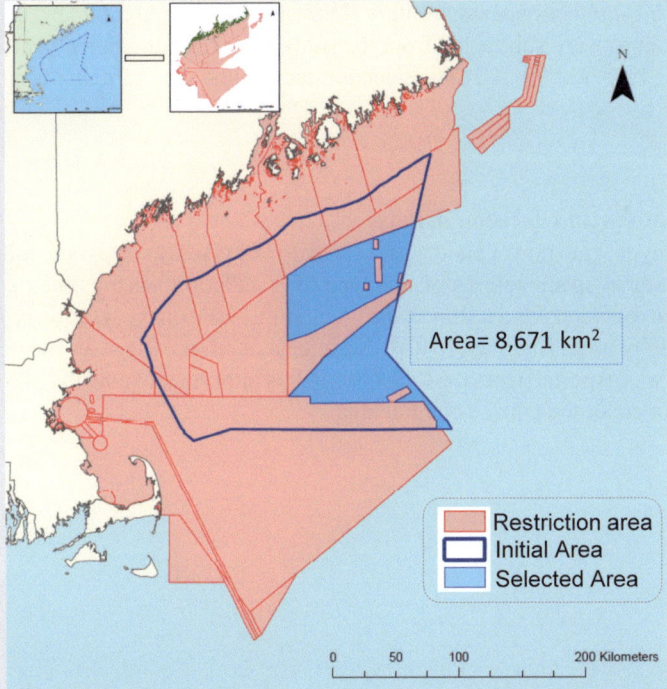
>
> **Fig. 4.2** Thematic layer of available locations without restrictive factors. Potential area. *Source* Gil-García et al. [5]

Table 4.1 Selection factors and data sources. *Source* Gil-García et al. [7]

Selection factor	Data source
Wind speed [m s^{-1}] Bathymetry [m]	Spatial layer
Wave height [m] Water quality [g/L]	CSV data and transformed into a spatial layer
PCC[a] distance [km] Ground distance [km] Distance to ports [km]	Calculated in GIS
CAPEX[b] [$] OPEX[c] [$] DECEX[d] [$]	Calculated externally and converted into a GIS layer

[a]Point of common coupling
[b]Capital expenditure
[c]Operational expenditure
[d]Decommissioning expenditure

Five layers of restrictive factors were identified: fishing zones, navigation routes, bio-protection areas, military zones, and submerged cables or pipelines. These zones were subtracted from the study area, resulting in a potential area of 8,671 km^2, see Fig. 4.2. The selection factors and data format are shown in Table 4.1.

Alternatives and decision matrix

A prototype wind power plant with a capacity of 1 GW was designed, following the technical specifications of this type of installation, such as the distances between rows and wind turbines, the orientation based on the wind rose, and the technology used. Eight alternatives were generated, see Fig. 4.3. The decision matrix was exported from the GIS system. It is important to note that the layout does not cover the entire potential area.

4.3 Application of GIS + MCDM

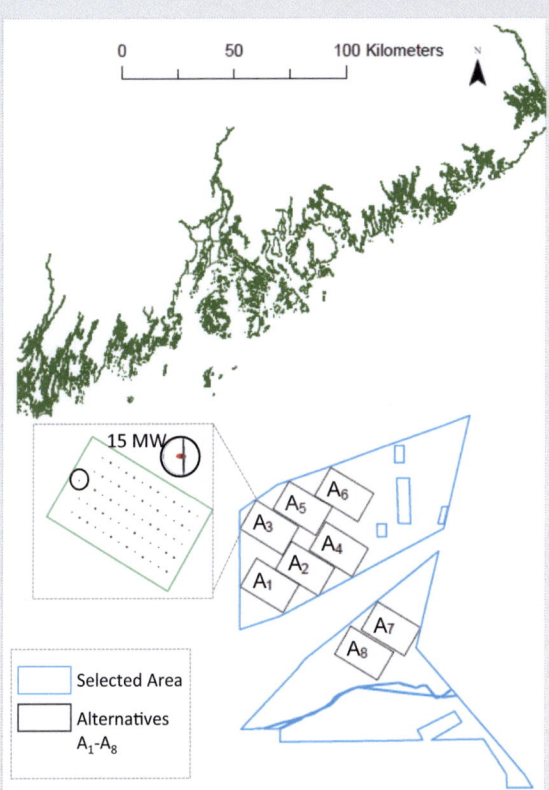

Fig. 4.3 Design of the alternatives. *Source* Gil-García et al. [5]

Weights and Ranking of alternatives

The weights of the selection factors were calculated using an expert matrix with AHP (see Sect. 2.2.1). The factors with the highest weight were wind speed, bathymetry, and distance to the connection point, with 33%, 22%, and 12%, respectively. The ranking of the alternatives was evaluated with a combination of the TOPSIS (see Sect. 2.3.2) and fuzzy (see Sect. 2.4) methods, both agreeing that the best options were A_3 and A_5. The final selection was made by calculating the specific electrical energy of each alternative using the WAsP software, determining that A_3 was the best option, with a net generation of 5,302 GW h/year, wake losses of 5.18%, and an LCOE of 100.4 $ MW h. The authors also performed a sensitivity analysis considering the annual generation, CAPEX and OPEX indicators, see Fig. 4.4.

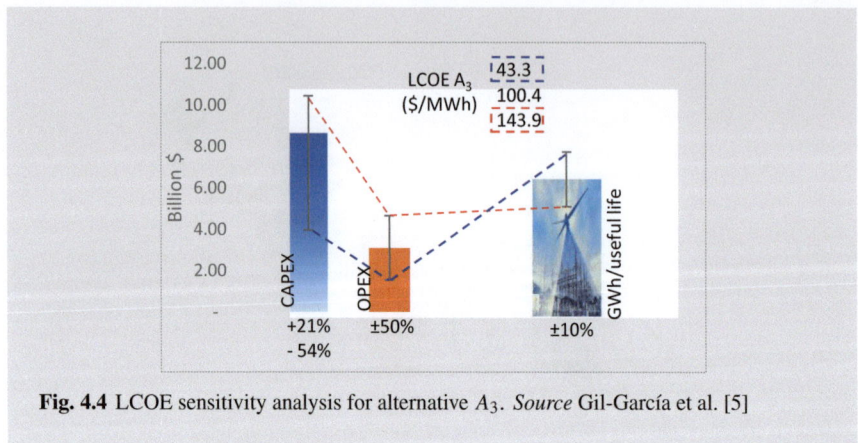

Fig. 4.4 LCOE sensitivity analysis for alternative A_3. *Source* Gil-García et al. [5]

4.3.1.2 Case Study: Spain

Gil-García et al. [8] presented a combined approach based on GIS and MCDM methodology to optimize the selection of offshore wind power plant locations. The methodology includes net annual electricity production as a key selection factor in the decision-making process, previously optimized in relation to the wake effect. The selection of potential locations considers criteria such as technical feasibility, environmental impact, economic viability and power generation potential. This MCDM-GIS approach was applied in a case study along the Spanish coast, considering 92 initial alternatives. The results indicated that the offshore wind energy targets set for Spain in 2030 and 2050 should be re-evaluated, as the optimal utilization of the available area represents only 16% of the total potential area. A suitability map was generated by integrating all relevant map layers and their respective zones of influence.

> **Study area, restrictive and selection factors, potential area**
> Unlike other countries, Spain has a well-defined maritime spatial planning (Spanish Maritime Spatial Planning, SMSP) area, the results of which are visible online for free through the "InfoMar" spatial tool [1].
> The restrictive factors for establishing areas with high offshore wind potential take into account economic, environmental, and social objectives. The main criteria are: protection of biodiversity, gravel deposits, protection of cultural heritage, research and development areas, national defence, safety of navigation, and areas for other economic activities (aquaculture, port activities, etc.).

4.3 Application of GIS + MCDM

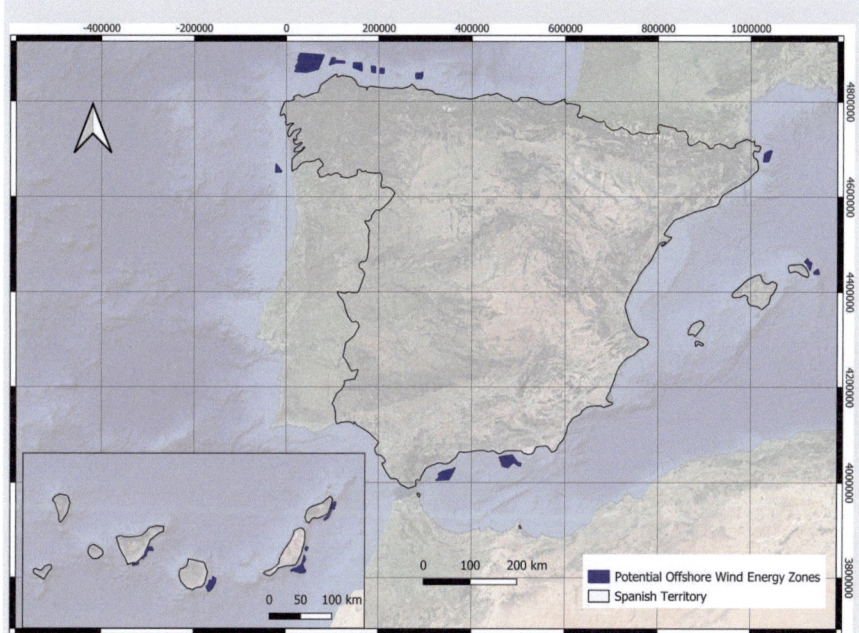

Fig. 4.5 Potential offshore wind Energy zones. *Source* Gil-García et al. [8]

The spatial tool obtained an area of 5,140 km² with high offshore wind potential, by differentiating the study area and the restrictive factors, see Fig. 4.5.

Alternatives and decision matrix

Two prototype installations are defined:

- **Prototype 1**: 450 MW of nominal power, and is composed of 30 wind turbines.
- **Prototype 2**: 225 MW of nominal power, and is composed of 15 wind turbines.

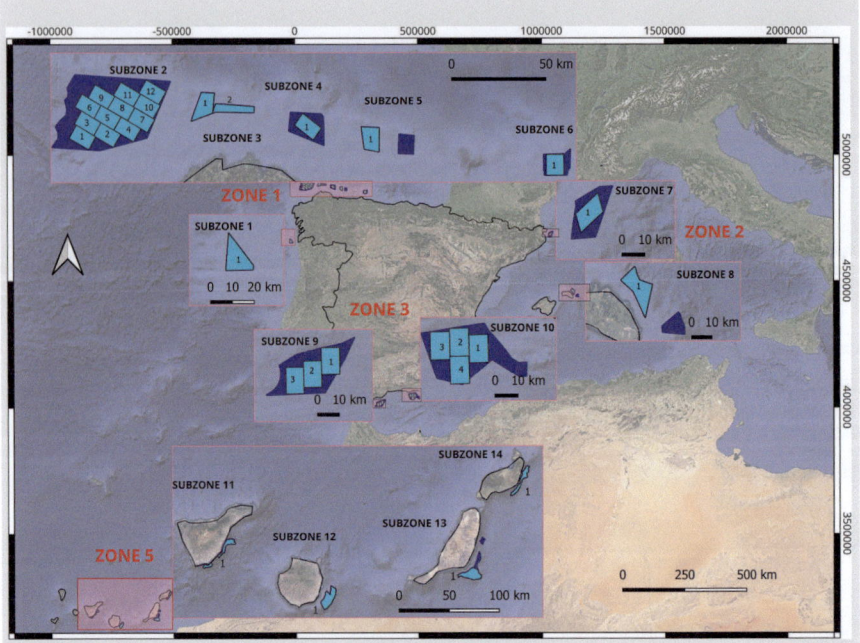

Fig. 4.6 Alternatives for prototype 1. *Source* Gil-García et al. [8]

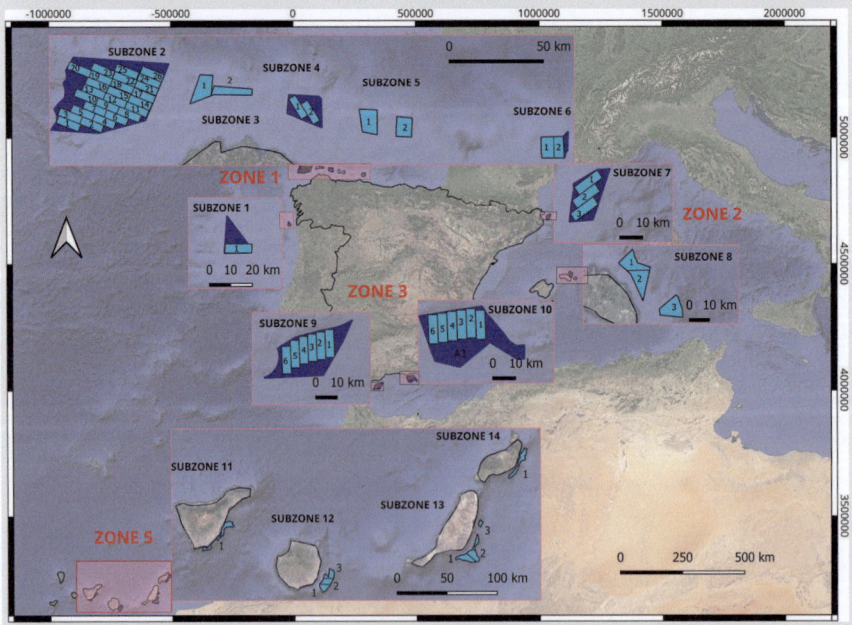

Fig. 4.7 Alternatives for prototype 2. *Source* Gil-García et al. [8]

Each prototype is distributed across the different potential zones, according to the predominant direction of the energy rose studied for each specific zone, obtaining 92 alternatives, see Figs. 4.6 and 4.7.

The selection factors and data format are shown in Table 4.2.

Table 4.2 Selection factors and data sources. *Source* Gil-García et al. [8]

Selection factor	Data source
Wind speed [m s^{-1}]	Calculated externally and converted into a GIS layer
Bathymetry [m]	Spatial layer
Wave height [m]	CSV data, R and transformed into a spatial layer
Area [km^2]	Calculated in GIS
Distance to ports [km]	
PCC^a distance [km]	
Ground distance [km]	
Net electricity production [GW h]	Calculated externally and converted into a GIS layer
CAPEX^b [$]	
OPEX^c [$]	

^{a}Point of common coupling
^{b}Capital expenditure
^{c}Operational expenditure

A key aspect of this study is that net electricity production was calculated for each of the 92 alternatives, optimizing it based on the wake effect. If the wake losses in the turbine layout of an alternative exceed 6%, a new iteration was carried out. Various specialists in the wind sector and technical design programmes were used in this process.

Weights and ranking of alternatives

The weights of the selection factors are determined by the AHP (see Sect. 2.2.1) and the Entropy (see Sect. 2.2.2) methods, and then aggregated through the Compromised method (see Sect. 2.2.3). The three factors with the highest weights were net electricity production, CAPEX, and bathymetry, with 52.3, 19.7 and 9.6%, respectively. The ranking of alternatives was determined by the TOPSIS (see Sect. 2.3.2) and VIKOR (see Sect. 2.3.3) methods, resulting in the first seven alternatives being the same in both methods although in different positions.

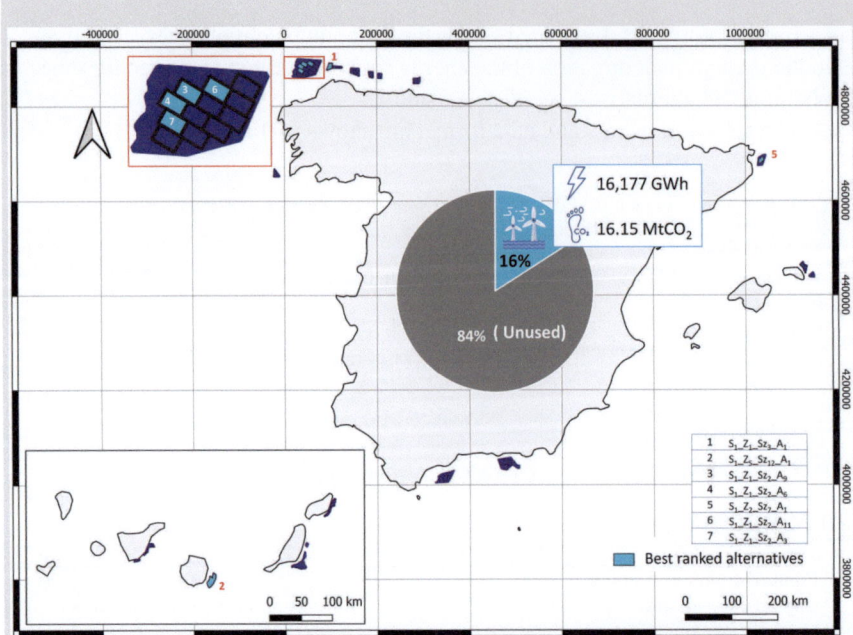

Fig. 4.8 Optimal alternatives according to the objectives of the 2030 Spain roadmap. *Source* Gil-García et al. [8]

The authors analysed the results with Spain's 2030 targets of reaching an installed capacity of 3 GW of offshore wind energy. To achieve this target, the first seven alternatives obtained by both methods were sufficient, which added up to an installed capacity of 3.15 GW. These seven alternatives would occupy an optimal surface area of 829 km^2, compared to the 5,140 km^2 of available area, leaving 4,311 km^2 unused. Therefore, it can be concluded that the 3 GW target is insufficient for Spain. Given the potential area, the roadmap should be reconsidered, see Fig. 4.8.

4.3.2 Onshore Wind Energy

4.3.2.1 Case Study: Balikesir, Turkey

Yildiz [25] proposed a GIS-based spatial MCDM approach to identify suitable sites for wind power plants in Balikesir, Turkey. Topographic, structural, climatic, and environmental criteria, such as wind speed, slope, and proximity to key infrastructures, were analysed. The weights of the criteria, obtained through the AHP process, were used to generate a suitability map. The results showed that 2.34% of the poten-

4.3 Application of GIS + MCDM

tial area was the most suitable for wind power plants. The criterion of proximity to seaports was introduced for the first time for onshore wind power plants.

Study area, restrictive and selection factors, potential area

The study area is Balikesir, located in the Marmara region of Turkey. The area of Balikesir is approximately 14,500 km². The exclusion factors, areas that are not suitable for onshore wind power generation due to legal and physical restrictions, were:

- Wind speed (at 100 m) < 5 m s^{-1}
- Land cover type (artificial surfaces): Wetlands, Water bodies
- Slope of terrain >30 %
- Distance from substations and electricity grid <250 m
- Distance from road networks, settlements, and fault lines <500 m
- Distance from seaports $>100,000$ m

The selection factors were determined from these same factors, and the authors defined different class ranges for each of them according to the literature. For example, in the case of the wind speed selection factor, the following ranges were defined: 5–6, 6–7, 7–7.5, 7.5–8 and >8 [m s^{-1}]. All factors were obtained in spatial layers and processed in a GIS. The restricted areas constituted approximately 15% of the study area, equivalent to 2,200.56 km².

Alternatives and decision matrix

Each range of selection factors was assigned a score from 0 to 5, where 0 corresponded to the restriction factors and the rest represented the suitability scores, with 5 being the highest. With this classification and assignment of scores, the information was processed in a GIS, generating scored criteria maps.

Weights and ranking of alternatives

The AHP method (see Sect. 2.2.1) was used to determine the weights of the selection factors with respect to the judgments of experts and academics with experience in the wind energy sector. Wind speed scored 40.8% followed by land cover type with 13.9%; the lowest weighted factor was distance from seaports with 3%.

The wind power plant suitability map was generated using the AHP weights and the produced criteria maps. In this map, 2.34% (344.32 km²) of the areas included in the analysis were in the most suitable class (5), and 9.34% (1,373.40 km²) were in the next most suitable class (4).

It is important to note that, as this was a suitability map, specific alternatives were not identified, but areas. It would be interesting, in the future, to study the best area considering technical and economic factors under different installed capacity scenarios.

4.3.2.2 Case Study: Región de Murcia, Spain

Sánchez-Lozano et al. [23] addressed the selection of locations for onshore wind power plants by combining MCDM methods with fuzzy approaches, which allow the inclusion of both numerical and qualitative criteria. They used the fuzzy AHP to calculate the weights of the criteria and the fuzzy TOPSIS technique to evaluate the alternatives. A GIS provided the criteria database, transformed into a fuzzy decision matrix.

Study area, restrictive and selection factors, potential area

The study area, the coastline of the Region of Murcia (southeastern Spain), has an area of $4{,}456.59\,\mathrm{km}^2$. The restrictive factors were imported from the web map services of the national and regional administrations, in the form of spatial layers to the GIS program. These factors included:

- Urban lands and protected and undeveloped lands
- Areas of high landscape value, water infrastructure, military zones and cattle trails
- Archaeological sites, palaeontological sites, and cultural heritage
- Roads and railroad network
- Community interest sites
- Areas of special protection for birds
- Watercourses and streams of the Mediterranean coast
- Cadastral municipalities of the coastline of Murcia

Once the restricted areas of the study area were eliminated, a thematic layer that allowed the visualization of the locations suitable for onshore wind power plants was created, highlighting 33,290 available locations, which represents 19.94% of the study area (potential area of $888.75\,\mathrm{km}^2$).

The thematic layers that defined the selection factors were identified from the existing bibliography, and the spatial data were obtained through web map services of the national and regional administrations and private companies. The following have been taken into account (see Fig. 4.9):

4.3 Application of GIS + MCDM

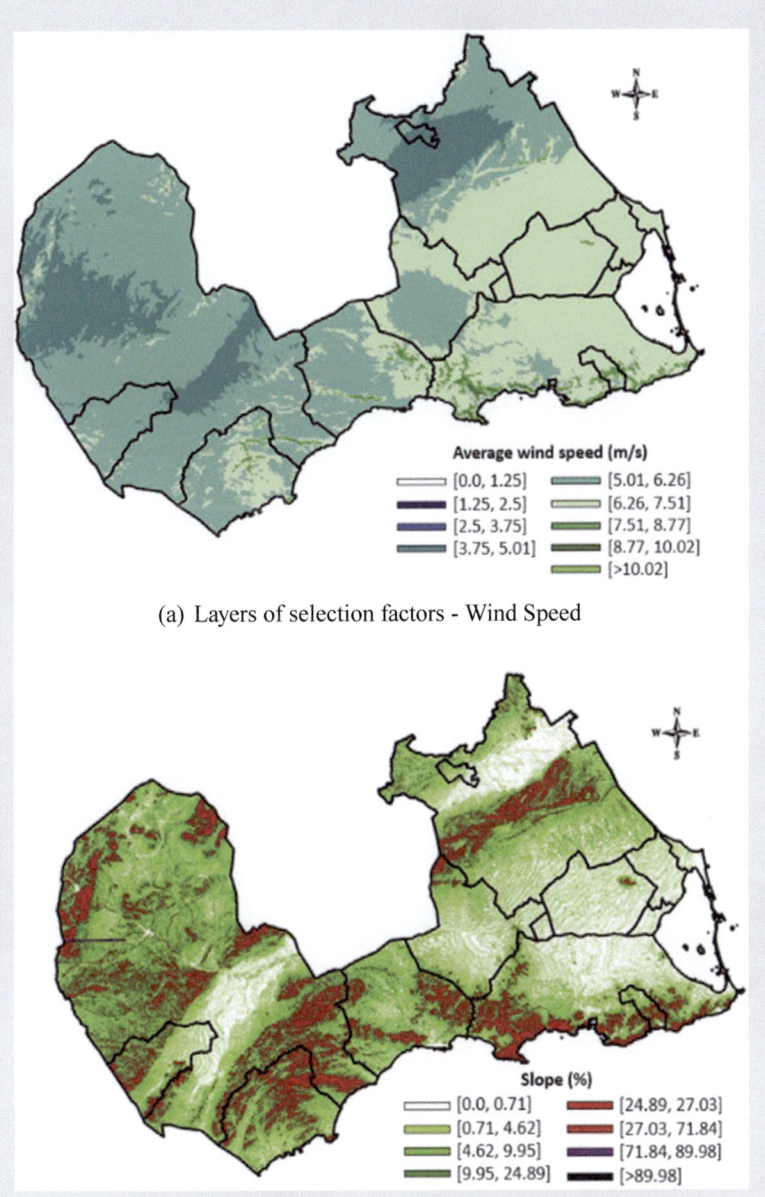

(a) Layers of selection factors - Wind Speed

(b) Criteria maps - Slope

Fig. 4.9 *Source* Sánchez-Lozano et al. [23]

- Agrological capacity [classes]
- Slope [%]
- Area [m^2]
- Distance to airports, roads, power lines, cities, electricity substations, mast [m]
- Average wind speed [m s^{-1}]

Alternatives and decision matrix

Once the different layers were incorporated, it was necessary to carry out editing tasks through the GIS (spatial link, filter, intersection, area, slope, and buffers) in order to obtain a global thematic layer (see Fig. 4.10). This layer provided the decision matrix with 33,290 alternatives to be evaluated, and was saved in Excel format to be able to apply the MCDM methods.

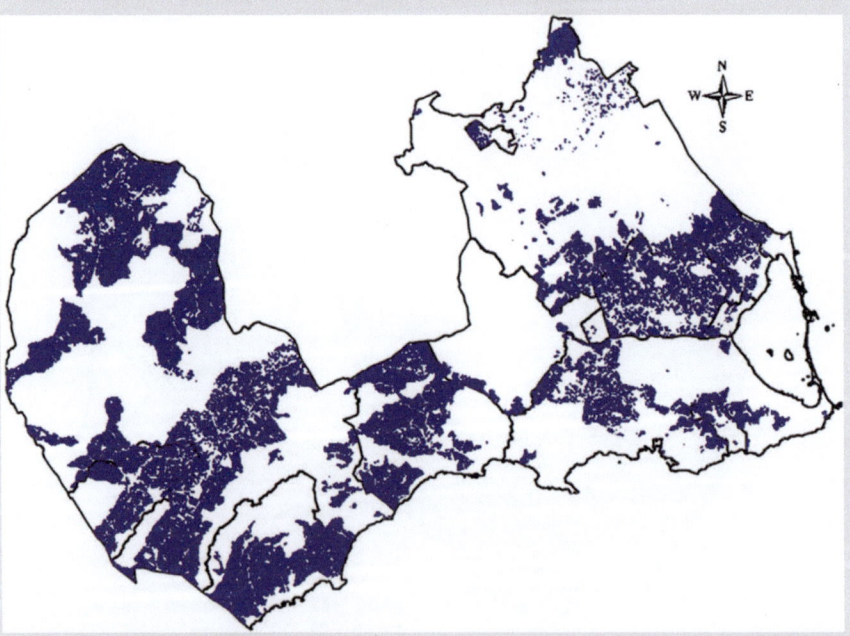

Fig. 4.10 Global thematic layer. Case Study: Región de Murcia, Spain. *Source* Sánchez-Lozano et al. [23]

Weights and ranking of alternatives

To determine the weights of the selection factors, a pairwise comparison was performed using the fuzzy AHP method, which is an extension of the traditional AHP (see Sect. 2.2.1), designed to address uncertainty and ambiguity in expert assessments. Instead of using exact or sharp values, it uses fuzzy logic to represent pairwise comparisons. In this logic, numerical values are replaced by fuzzy numbers or value intervals, allowing a more flexible representation of opinions and preferences. Using homogeneous aggregation, it is observed that the most important selection factor is wind speed, while the second most important is distance to cities. The least important were agricultural capacity and distance to airports.

The TOPSIS (see Sect. 2.3.2) method was used to obtain a measure of the effect produced by each alternative with respect to each selection factor. This method is especially useful for those problems where the values of the alternatives are not represented by the same units, as in this case study: linguistic labels, numerical values, and triangular fuzzy numbers. The authors solved the situation by adapting TOPSIS to the operations of fuzzy set theory. All the values obtained were classified with four sets of results based on their capacity to be used as a location for an onshore wind power plant (Excellent—Poor), see Fig. 4.11.

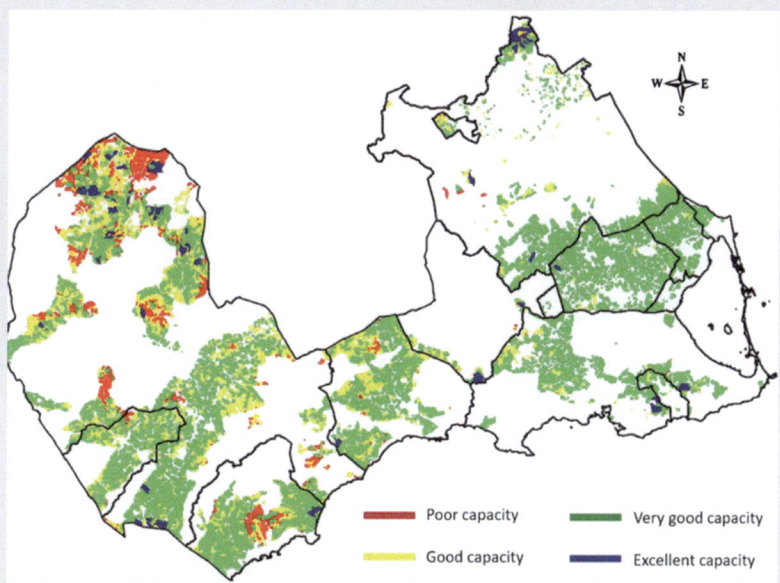

Fig. 4.11 Wind power plant suitability map. Case Study: Murcia, Spain. *Source* Sánchez-Lozano et al. [23]

4.3.3 Solar Photovoltaic

4.3.3.1 Case Study: Región de Murcia, Spain

Sánchez-Lozano et al. [24] combined GIS and MCDM to evaluate the optimal location of solar PV power plants in Cartagena, Región de Murcia, Spain. The GIS-MCDM combination allowed generating a detailed database and to simplify decision-making by applying multiple criteria. Two types of criteria were considered in GIS: restrictive criteria, which delimited the area excluding unsuitable zones according to legislation, and weighting criteria, which influenced the viability of the locations. Using the AHP method to weight the criteria and TOPSIS to evaluate the alternatives, optimal sites for the solar PV power plant were identified.

> **Study area, restrictive and selection factors, potential area**
> The study area is Cartagena, located in Región de Murcia, Spain. The average global annual radiation in most of its territory exceeds 5 kW/m^2 day. The restrictive factors were incorporated in spatial layers, highlighting the following:
>
> - Land classification
> - Infrastructure, military zones, and cattle trails
> - Cultural heritage
> - Palaeontological and archaeological sites
> - Water courses and streams
> - Areas of special protection for birds
> - Cadastral municipalities of Cartagena
> - Agrological capacity
> - Field orientation and slope
> - Potential solar radiation
> - Power lines
>
> The restrictive layers were treated in GIS, obtaining a layer with the appropriated areas (potential zones), which covered a total of 270.78 km^2, 25.50% of the studied area, see Fig. 4.12.

4.3 Application of GIS + MCDM

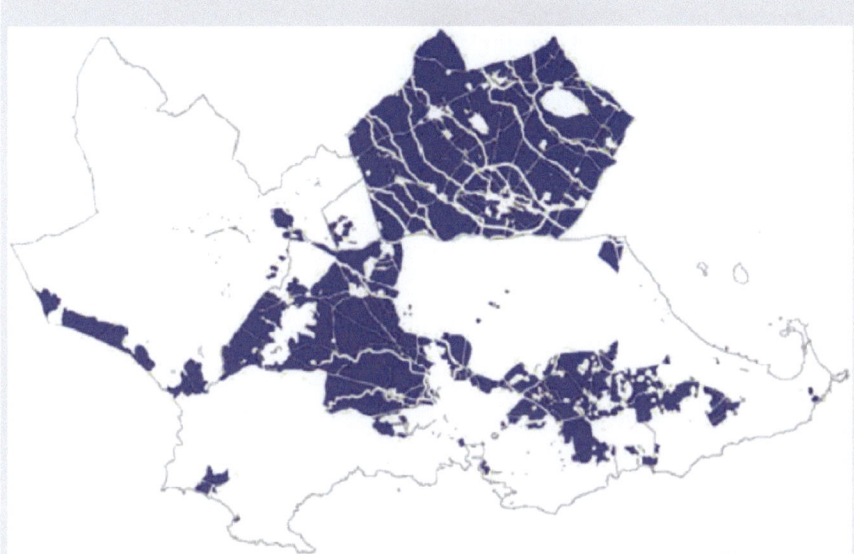

Fig. 4.12 Suitable areas. Case Study: Cartagena-Murcia, Spain. *Source* Sánchez-Lozano et al. [24]

From this layer, all built-up areas were removed and a prototype design of a solar plant with an area of $1{,}000\,m^2$ was created. The selection factors for establishing an order of alternatives were the following:

- Agrological capacity
- Land slope and orientation
- Land area
- Distance to villages, main roads, substations, and power lines
- Solar irradiation
- Average temperature

Alternatives and decision matrix

12,655 available plots were spatially identified. The GIS software generated a table according to the relational database model, which showed the thematic information represented in rows and columns, where the rows constituted the geographic objects. In this case, the geographic objects were the alternatives to be evaluated (parcels), and the columns (or fields) defined the thematic attributes or variables (cadastral information and selection factors).

Weights and ranking of alternatives

The weight of the selection factors was obtained with the AHP method (see Sect. 2.2.1), resulting in the location factors with the greatest weight (48.6%), and the environmental factors with the lowest weight (5.5%).

The TOPSIS method (see Sect. 2.3.2) was implemented in GIS using the "calculation tool". The results were sorted by intervals, each interval indicated the capacity of suitable plots: poor, good, very good, and excellent. The results showed that (see Fig. 4.13):

- 0.278% of the valid surface is not suitable for the implementation of solar PV power plants
- 0.773% has good load capacity
- 9.591% has very good capacity
- Finally, the remaining 3.206% is considered as excellent

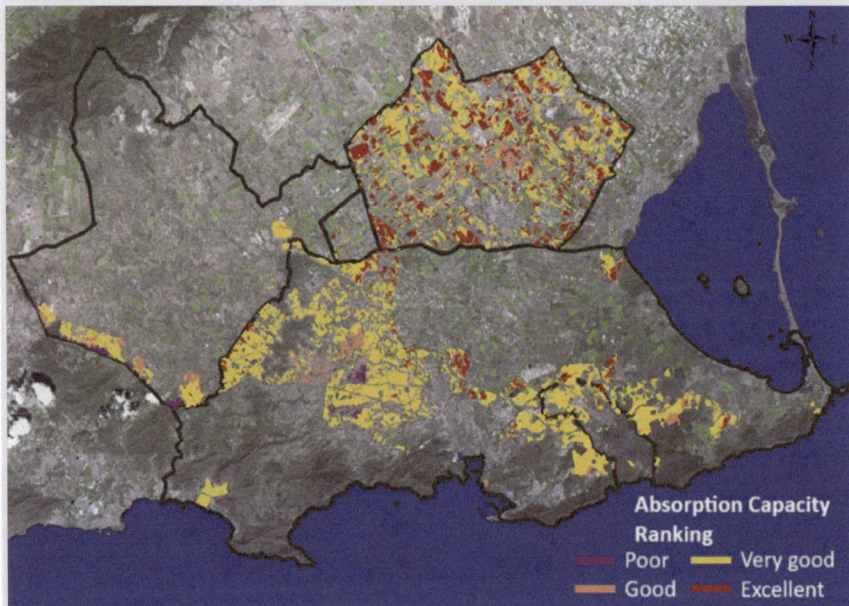

Fig. 4.13 Ranking. Case Study: Cartagena-Murcia, Spain. *Source* [24]

4.3.3.2 Case Study: Tunisia

Tunisia, with a high solar potential due to its climate and location, presents optimal conditions for solar PV projects. Rekik and El Alimi [19] used a MCDM and GIS approach to analyse land suitability and locate the best sites for installing solar PV power plants. Technical, economic, and environmental factors such as solar radiation, proximity to infrastructure, and land use were assessed using a fuzzy AHP method to weight the criteria. The results showed that 17.6% of the study area was viable for these projects, especially in the coastal and eastern areas, with a potential of 1,060 TW h/year.

Study area, restrictive and selection factors, potential area

The study area covered the territory of Tunisia, with a surface area of 163,610 km^2. The restrictive factors considered included:

- Distance from protected areas, the power grid, roads, and residential areas
- Terrain slope
- Land use

The restrictions were generated and added as a layer using Boolean algebra (0 or 1) within integrated GIS tools, resulting in a total excluded area equivalent to 12.61% of the entire study area.

The selection factors were selected according to the reviewed literature, categorized into climatic, accessibility, and topographic. They were:

- Global horizontal irradiance
- Temperatures
- Average cloudy days
- Proximity to grid
- Proximity to roads
- Proximity to urban areas
- Proximity to water resources
- Slope
- Aspect: referring to the geographic area (north, south, east, west, etc.)
- Land use
- Soil texture

Alternatives. Decision matrix

In this study, the spatial analysis was started by rescaling, resampling, and reclassifying the various input layers. Each layer was then adjusted using a fuzzy membership function within spatial analysis tools. To assign weights based on the relevance of each factor, the fuzzy AHP technique (see Sect. 2.2.1) was used. Subsequently, the final suitability map was generated using fuzzy overlay, based on the GIS-supported MCDM model. Finally, a minimum threshold of $1\,km^2$ was applied to identify the most suitable locations for installing large-scale PV power plants.

Weights and ranking of alternatives
The weighting of selection factors was determined using the fuzzy AHP method, based on criteria provided by a group of experts, who gave their opinion.

Solar radiation resource was the factor with the highest weight, with a relative value of 27.4%. The next most relevant selection factors were proximity to the electrical grid and transportation infrastructure, with weights of 17.9 and 11.7%, respectively. Ambient temperatures and slope also proved to be important factors, with relative weights of 10.6 and 8.5%. Other factors such as average cloudy days, orientation, distance to residential areas, and access to water resources scored between 4.1 and 5.8%. As for land use type and texture, they were considered less influential, with values of 2.5 and 1.1%, respectively.

The results of the site suitability assessment revealed that approximately $28,781\,km^2$ (17.6% of the total study area) were considered suitable for installing large-scale solar PV systems. The central, southwestern, southeastern, and eastern coastal areas of the country presented numerous locations that were particularly suitable for solar PV installations. In contrast, the northern areas, which contained the largest croplands and contiguous mountains, were considered unsuitable for constructing large-scale solar PV systems.

To improve clarity and provide more detailed information on the suitability of the sites, the authors categorized the results into three classes: highly suitable, suitable, and moderately suitable. According to this classification model method, it turned out that a total area of almost $5,251\,km^2$ (3.31% of the available surface area) presented a high potential for the development of solar PV energy.

4.3.4 Geothermal

4.3.4.1 Case Study: Continental Scale. EU Pan European Maps

Study area, restrictive and selection factors, potential area
The study area includes the 27 EU Member States plus 9 additional European countries, covering a population of approximately 500 million people. This region offers diverse geological, climatic, and environmental contexts that affect RES potential, particularly for shallow geothermal energy (SGE) systems. A GIS-based assessment was applied across this area, gathering indicators such as lithology, heating and cooling degree days, ground surface temperature, and environmental protections to create detailed suitability maps for renewable resource exploitation.

Several factors restrict renewable energy deployment in this study, particularly related to the technical and environmental viability of SGE systems. Exclusion criteria included:

- Geological factors affecting drilling (e.g., consolidation grade and aquifer productivity)
- Temperature requirements based on surface temperatures and heating/cooling demands
- Environmentally protected zones where energy extraction is restricted or prohibited due to conservation laws

These restrictive layers were processed in GIS to define potential areas for renewable energy projects. The restrictive zones cover about 15% of the study area, limiting exploitation in regions with unsuitable geological or climatic conditions and significant environmental protections.

Spatial overlay analysis
The thematic maps generated in this study offer valuable spatial information across key factors impacting shallow geothermal energy suitability. Overlaying these maps with population density maps provides additional insights into the alignment of energy resource availability with population needs. For instance, the thermal conductivity map (Fig. 4.14) highlights areas with higher conductivity values, indicating optimal zones for efficient SGE installation, which, when compared with high-density population areas, suggests regions where the energy impact could be maximized. Similarly, the ground surface temperature map (Fig. 4.15) allows identification of locations with ideal temperature ranges for SGE systems; when paired with population maps, this overlay suggests regions where demand and resource conditions are favourable.

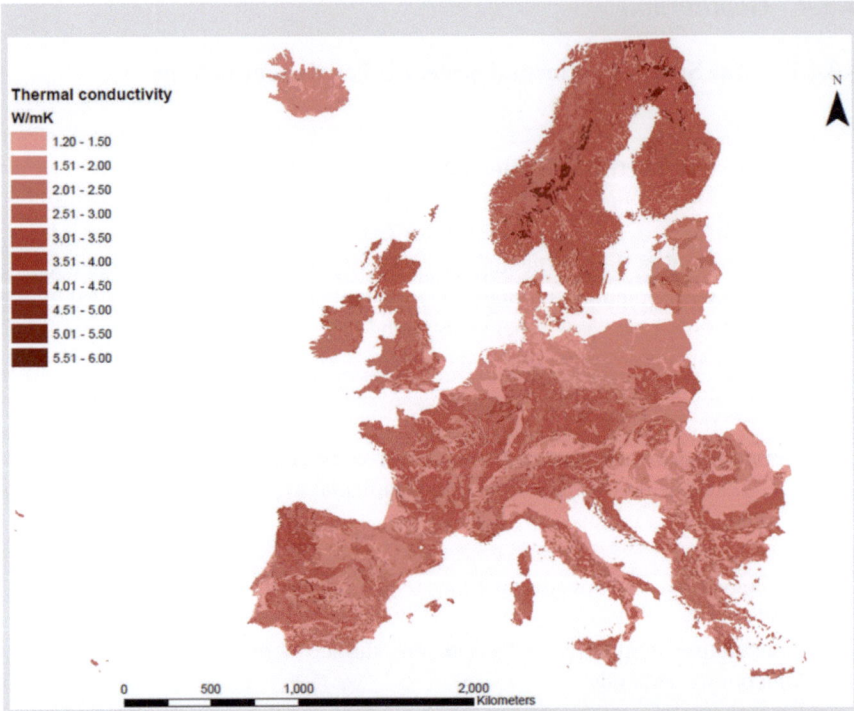

Fig. 4.14 Thermal conductivity thematic layer. Case Study: European scale. *Source* Ramos-Escudero et al. [18]

Additionally, the protected areas map (Fig. 4.16) reveals zones where renewable energy projects may face restrictions due to environmental protections. The annual thermal amplitude (Fig. 4.17) is of great importance when it comes to determining the suitability of providing space heating and/or cooling with GSHPs compared to more conventional air-source heat pumps (ASHPs). Compared to ground, ambient air shows a wider variation in temperature all year-round and on a daily basis, which reduces the COP. Mapping this alongside population density informs decision-makers about areas where renewable installations are feasible and where legal constraints exist, helping to prioritize locations with minimal regulatory hurdles. By integrating these spatial overlays, energy policymakers can make more informed decisions, optimizing RES infrastructure to benefit the greatest number of people while considering environmental and technical factors.

4.3 Application of GIS + MCDM

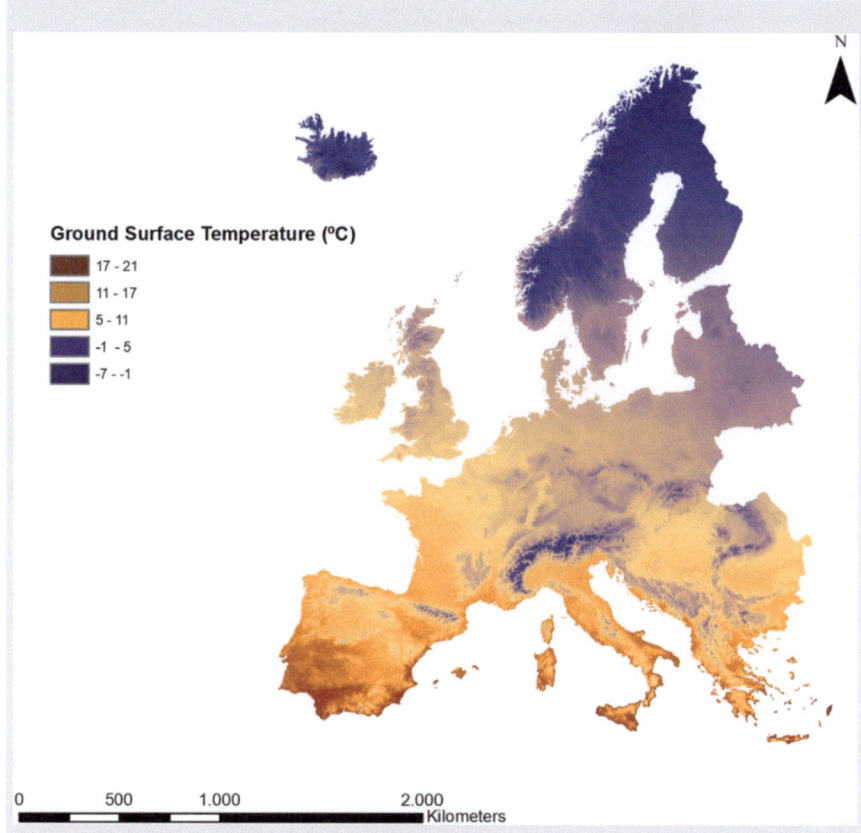

Fig. 4.15 Ground surface temperature thematic layer. Case Study: European scale. *Source* Ramos-Escudero et al. [18]

Alternatives and decision matrix

For this work, as SGE could be exploited almost everywhere, is the suitability level of exploiting SGE by GHSP in a certain area, rather than the optimal placement, the goal that could be reached.

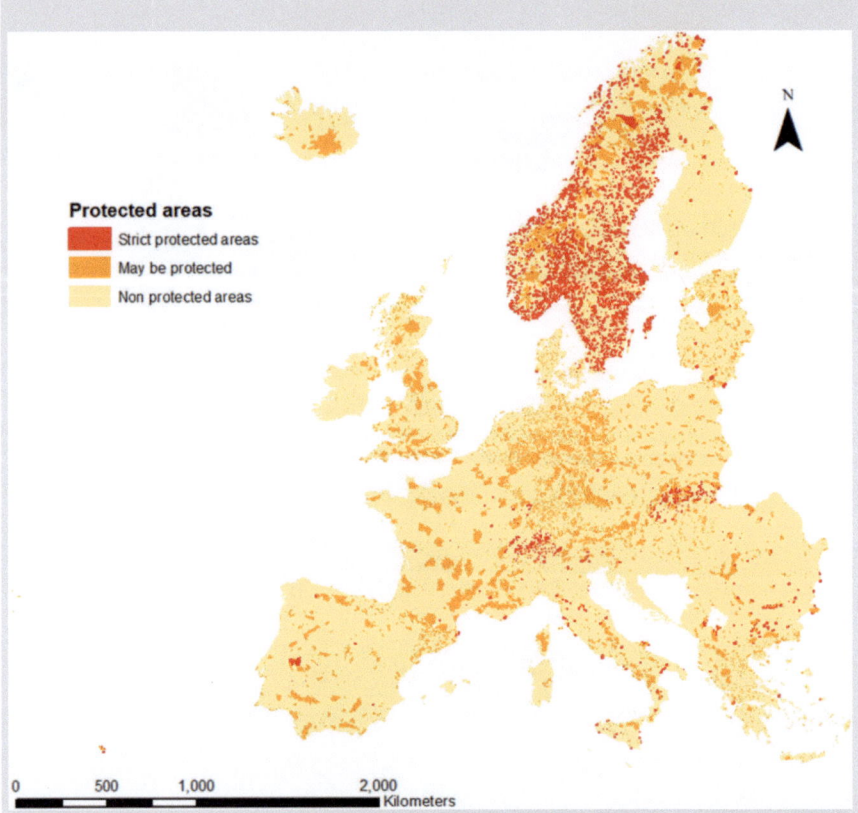

Fig. 4.16 Protected areas thematic layer. Case Study: European scale. *Source* Ramos-Escudero et al. [18]

The criteria are weighted, according to the understanding of hypothetical experts, based on their influence to determine the areas where GSHP will perform with the best results. To be able to compare one criterion to another, a normalization of the data was carried out, where quantitative data were also translated into qualitative. In this process, each criteria value was first categorized into three groups according to their suitability to reach the goal and a score was assigned to the categories: 1 is for the category considered acceptable, 2 is good, and 3 is excellent (see Fig. 4.19 and Table 4.3).

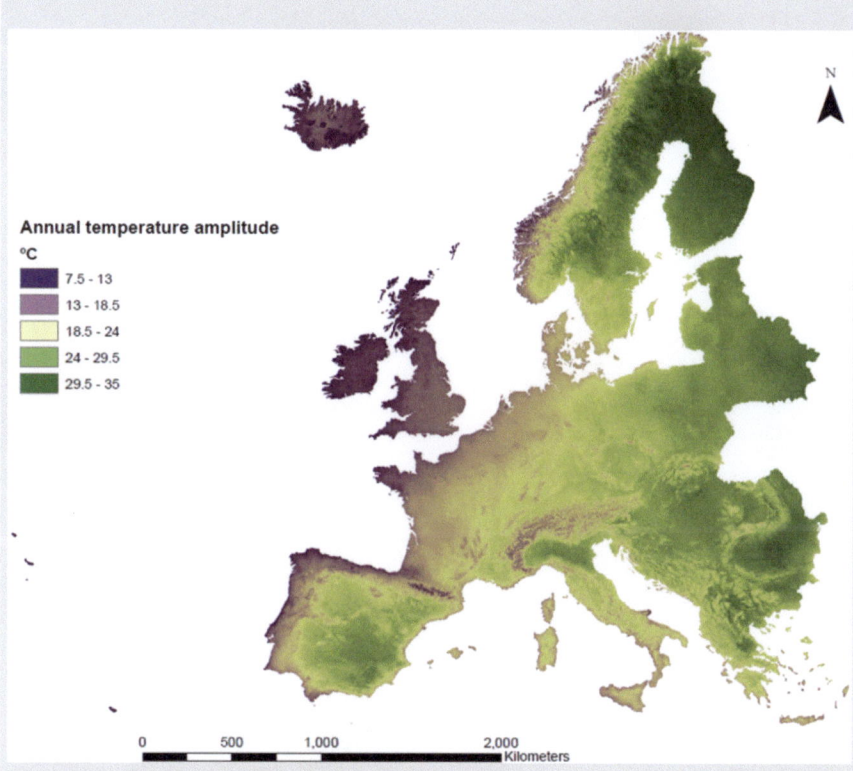

Fig. 4.17 Annual thermal amplitude thematic layer. Case Study: European scale. *Source* Ramos-Escudero et al. [18]

Weights. Ranking of alternatives

From the Pan-European maps, a regional MCDM-GIS was developed in Southeast Spain so that only extraction of the maps for this area was considered (see Fig. 4.18. The weighted sum in GIS was made using the "map algebra" function. It was assumed that $w1$, $w2$, $w3$, $w4$, $w5$, and $w6$ were the weights for every criterion and $S1$ (thermal conductivity), $S2$ (heating + coolling Degree Days), $S3$ (rock consolidation grade), $S4$ (ground surface temperature), $S5$ (protected areas) and $S6$ (annual temperature amplitude) are the scores previously assigned. Results showed that the suitability level of this region for SGE systems is:

- Acceptable around 20% of the territory
- Good in 40% of the territory
- Excellent in around 20% of the territory

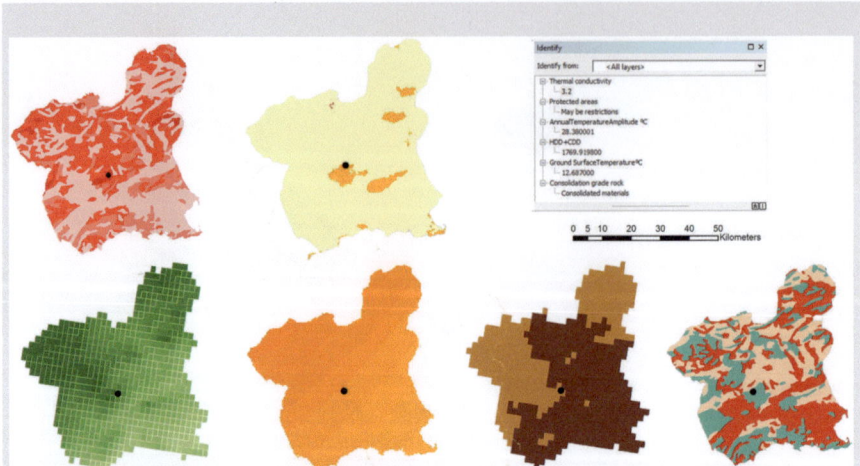

Fig. 4.18 Suitability map. Case Study: European scale to regional scale. *Source* Ramos-Escudero et al. [18]

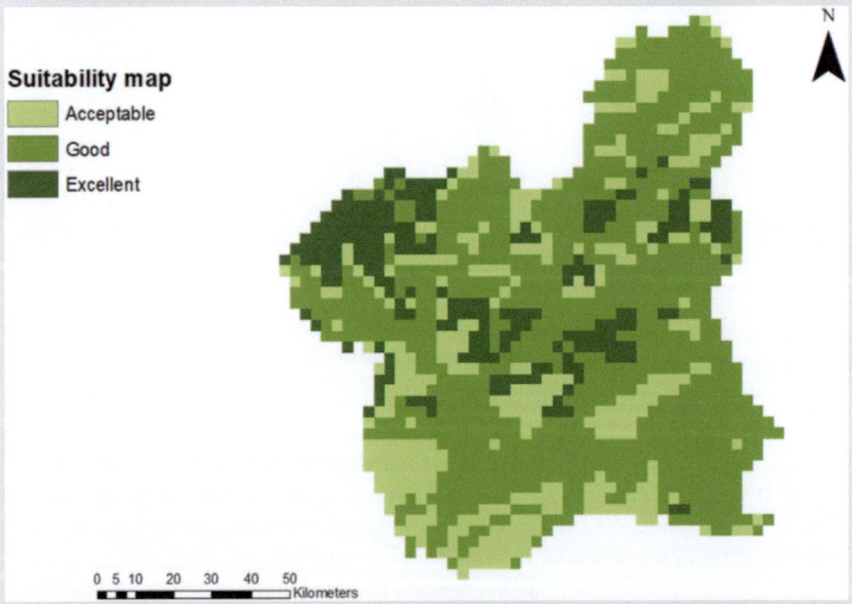

Fig. 4.19 Suitability map. Case Study: European scale to regional scale. *Source* Ramos-Escudero et al. [18]

Table 4.3 Weight of criteria and factors applying MCDM and GIS in Spain at a regional scale

Order	Criteria	Weight	Suitability label		
			(1) Acceptable	2 (Good)	3 (Excellent)
1st	Thermal conductivity	0.6	1.2–2.5	2.51–4	4.01–6
2nd	HDD-CDD	0.1	8,000–2,000	2,001–4,300	4,301–7,230
3rd	Consolidation grade rock	0.1	Consolidated	Partly cons.	Unconsolidated
4th	Ground Surface Temp.	0.09	(−7.2)–2	2–18	18–21
5th	Protected areas	0.07		May be restrictions	Non protected areas
6th	Annual Temp. Amplitude	0.04	7.5–18	18–24	24–35

4.3.4.2 Case Study: Madrid, Spain

Study area, restrictive and selection factors, potential area

"Ciudad Lineal" is a neighbourhood in Madrid, Spain, known for its ample outdoor spaces and high population density. Covering an area of 11.4 km^2 within the 604.3 km^2 of Madrid, it has a density population of 87 inhabitants per hectare, exceeding the municipal average of 52 inhabitants per hectare. This neighbourhood not only highlights the significance of urbane in densely populated areas, but also serves as an optimal case study for analysing urban systems. Additionally, Madrid accounts for 15% of Spain's total population. The aim was to analyse a PV-GSHP hybrid system.

The restriction factors were:

- Open public transport infrastructure
- Building inner space and safety distance from both buildings and infrastructure (3 m)
- Main roads and highways

The selecting factors were:

- Building energy demand
- Number of borehole heat exchangers
- Electricity production from the PV panels
- Coupled system cost
- Emission saved

Alternatives and decision matrix

The decision matrix consisted of five criteria and 2,733 alternatives, representing the buildings located in the neighborhood (Fig. 4.20). The criteria measured both the capacity of the area to provide the thermal needs through GSHP in open areas such as parks, backyards, etc., and the capacity of the rooftops to provide space to install the PV panels needs. In addition, the economic and environmental costs were also assessed (Figs. 4.21 and 4.22).

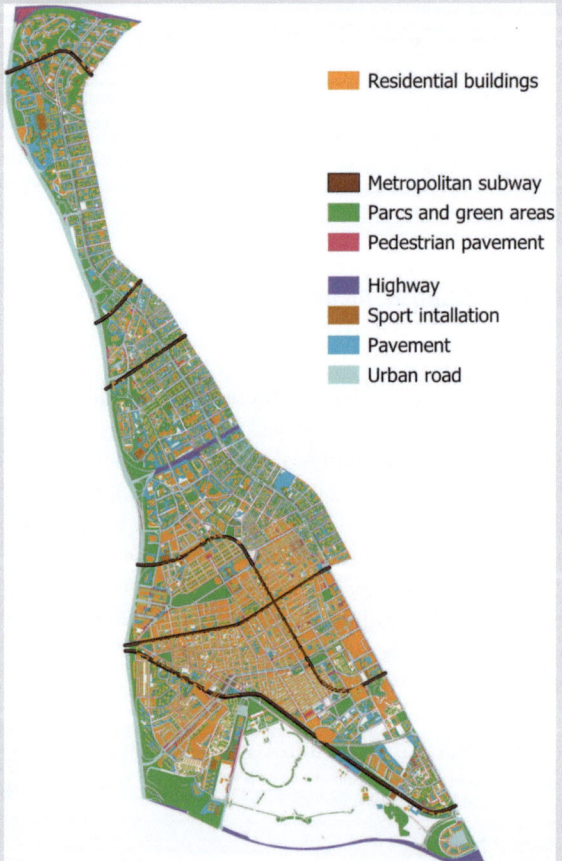

Fig. 4.20 Residential buildings alternatives in Madrid. *Source* Ramos-Escudero et al. [17]

4.3 Application of GIS + MCDM

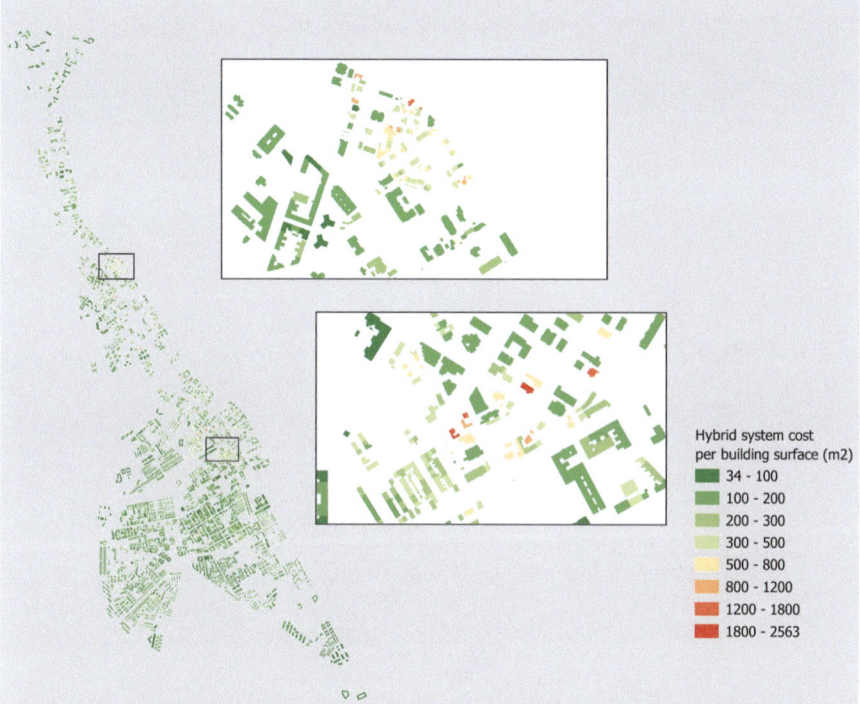

Fig. 4.21 Residential buildings alternatives in Madrid. *Source* Ramos-Escudero et al. [17]

Weights. Ranking of alternatives

Using the Entropy method (see Sect. 2.2.2), greenhouse gas (GHG) emissions' savings emerged as the most influential criterion with a weight of 17.11%, aligning strongly with Sustainable Development Goal (SDG) 13: Climate Action. Other criteria, including PV energy production (16.88%), heat island effect (16.55%), BHE+PV cost (16.52%), building age (16.47%), and energy demand (16.46%), had similar but slightly lower weights.

The ranking of buildings, shown in Fig. 4.23, are determined using the VIKOR method (see Sect. 2.3.3). It categorized their suitability for PV-geothermal systems from highly optimal (dark green) to less optimal (dark red). The northern and central regions, represented by dark green, were identified as the most favorable zones (top-ranked 1–250) for deployment. In contrast, the southern areas exhibited lower suitability, with a mix of yellow, orange, and red shades. Based on the sorting of alternatives, they were ranked according to their indices, with lower indices indicating better performance.

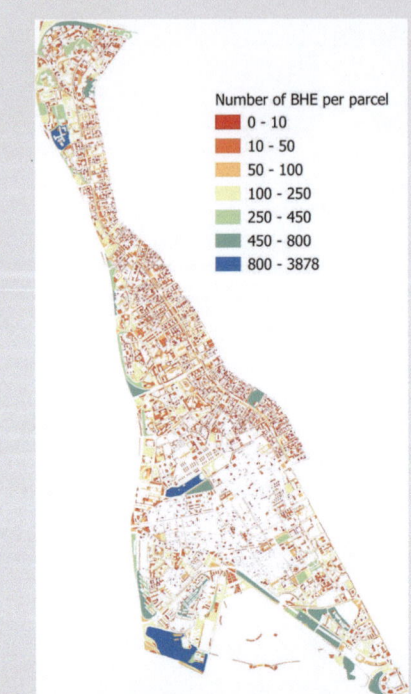

Fig. 4.22 Residential buildings alternatives in Madrid. *Source* Ramos-Escudero et al. [17]

Ranking results allowed for prioritization of budget forecast scenarios: four investment scenarios were developed, varying by budget and public investment percentages, to prioritize projects based on energy and CO_2 savings. Lower-budget scenarios focused on high-impact, costlier projects, while higher budgets enabled to address more buildings with lower unit costs. Early years prioritize projects with significant environmental benefits, while later years implement lower-cost interventions, increasing the number of annual projects. The methodology ensures maximum reductions in energy consumption and CO_2 emissions, balancing sustainability goals with financial constraints. Over 10 years, this approach optimized resource allocation and prioritized impactful projects to achieve substantial environmental and energy-saving improvements.

4.3 Application of GIS + MCDM 97

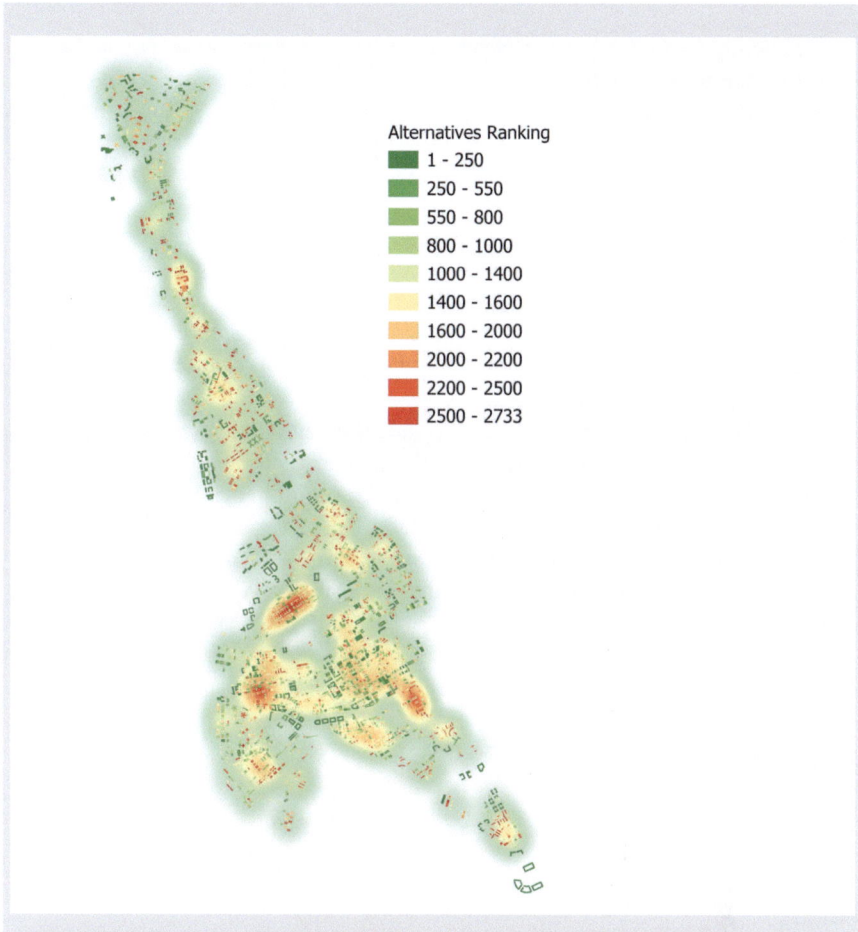

Fig. 4.23 Residential buildings alternatives in Madrid. *Source* Ramos-Escudero et al. [17]

4.3.4.3 Case Study: Locations for PV-Geothermal Systems for Industrial Areas in Spain

Study area, restrictive and selection factors, potential area
Spain (along with Portugal) accounts for the highest yearly radiation rates, which energy is mainly used to produce electricity. Moreover, in Spain, 31% of the total energy is consumed by industry, eminently consumed in industrial parks, considered as energy-consumer hotspots. Therefore, a hybrid PV-geothermal system is proposed as a more efficient system to produce heat and cold from RES for the industry, both for industrial processes and heating

and cooling (H&C). However, Spain shows very different orography and climate conditions along its territory, causing different performance ratios of the proposed hybrid systems. Among the most influential factors affecting these systems' environmental and economic performance, are the yearly solar irradiation, the energy demand, and the geothermal resource potential that differ in a spatial component.

In this work, a MCDM tool was created and applied to spatially to identify optimal locations for PV-geothermal systems for H&C in industrial areas. No restriction factors were considered, whereas the selection factors were selected according to the reviewed literature, categorized into resource, environmental, and technical:

- Shallow geothermal resource and technical potential factors: thermal conductivity and maximum thermal energy to be extracted from the ground.
- Solar resource and technical potential factor: solar direct normal irradiation and maximum electricity to be generated by the PV panels.
- PV/Geothermal surface footprint: how many m^2 of solar panels are needed to provide the energy demanded by the GSHP to satisfy the energy demand.
- Emissions saved.

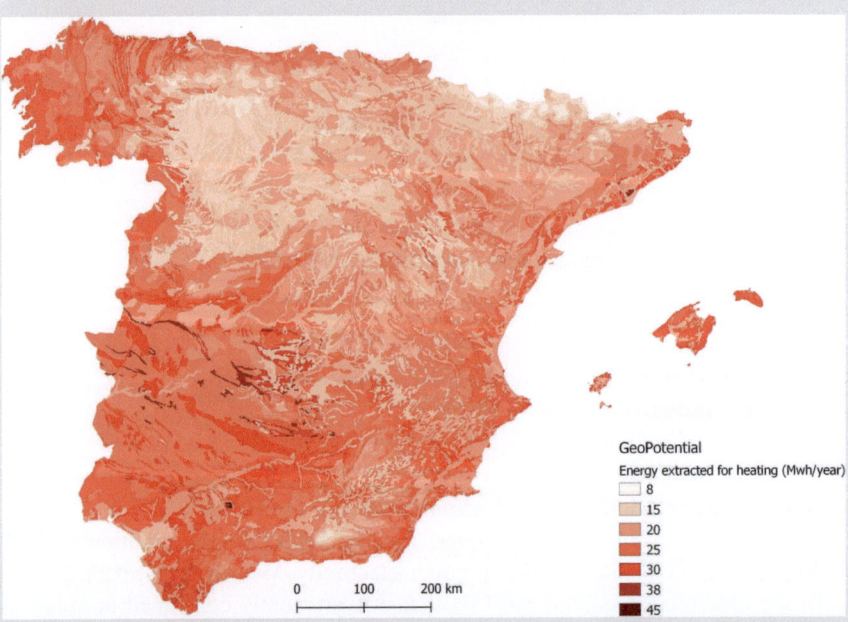

Fig. 4.24 Energy that can be annually extracted from the ground to provide space heating. *Source* Escudero et al. [4]

4.3 Application of GIS + MCDM

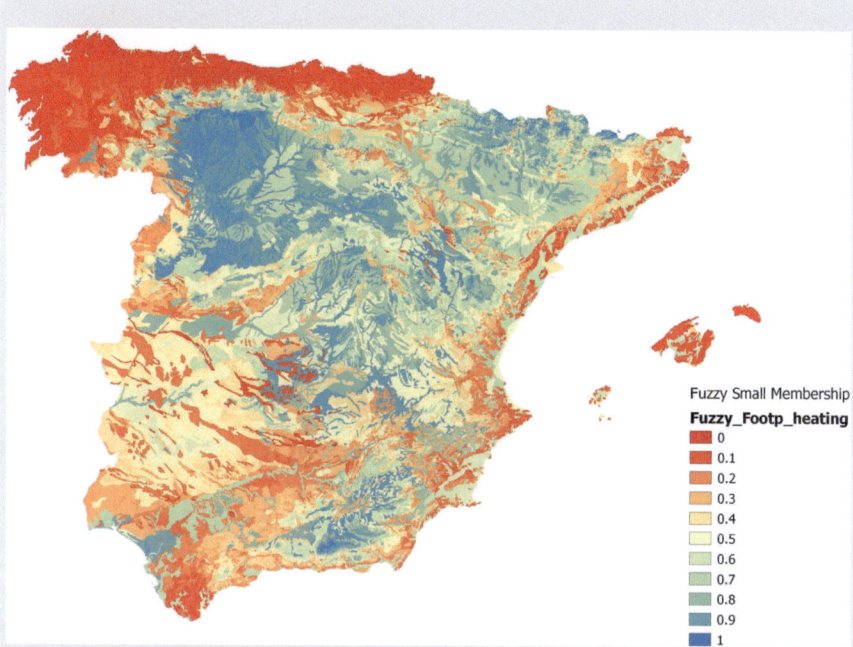

Fig. 4.25 Ratio PV-Geothermal. *Source* Escudero et al. [4]

From the literature, it was known that the geothermal resource was one of the most influencing criteria, see Fig. 4.24. Moreover, the PV/geothermal ratio map (Fig. 4.25) gave insight over the spatial suitability of the hybrid system, based on the solar and geothermal resource throughout the territory of Spain.

Alternatives and decision matrix

The spatial analysis began by calculating various factors, followed by adjusting each layer using a fuzzy membership function within the spatial analysis tools in a GIS environment. While no specific weights were assigned in this case study, the relative importance of the factors was either amplified or reduced to influence the final decision. The final suitability map was then generated using a fuzzy logic approach within a GIS-supported MCDM model. To refine the results, a minimum threshold of 1 km^2 was applied to identify the most suitable locations for installing PV-geothermal systems.

GIS fuzzy membership functions normalize and transform the input values into a standardized range, typically between 0 and 1 or, in some cases, between 1 and 10. This normalization process facilitates the comparison and integration of different datasets by expressing the degree of membership or suitability in a consistent and dimensionless format (Fig. 4.26).

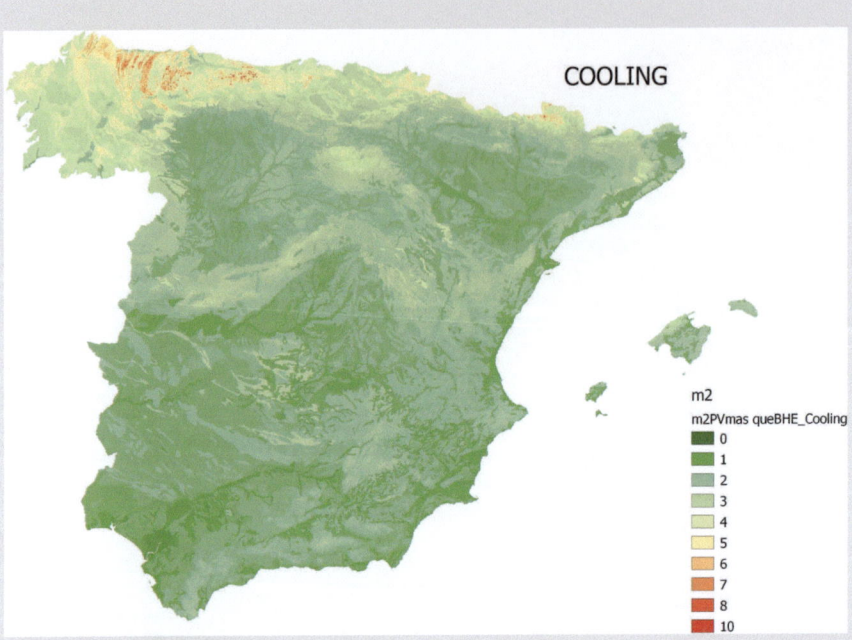

Fig. 4.26 Energy that can be annually extracted from the ground to provide space heating. *Source* Escudero et al. [4]

Weights and ranking of alternatives

In the MCDM process, fuzzy membership functions are applied to evaluate factors based on their relevance to the final decision, either maximizing or minimizing their impact. For example, the PV-geothermal surface footprint uses a membership function to minimize impact, where smaller values have a membership closer to 1, as lower footprint is preferred. Conversely, the GHG savings potential is maximized, with larger values having a membership closer to 1, reflecting the preference for higher savings. This normalization ensures consistent integration of datasets, enabling a GIS-supported MCDM model to effectively balance and prioritize factors for decision-making.

The results of the site suitability assessment revealed that approximately 70% of the total study area were considered with "Very low potential" for installing solar PV systems for industrial areas; from "low" to "very high" varied similarly in numbers (around 7%), although "very high" potential areas are concentrated in the Southwest of Spain. This is mainly due to the space available for the utilities and the strong resource potential of these areas.

4.4 Conclusion

As highlighted throughout this chapter, the integration of GIS and MCDM provides a robust framework for addressing the complexities of RES planning. GIS facilitates the visualization and spatial analysis of critical geospatial data, while MCDM allows for a structured evaluation of diverse factors, including technical, economic, environmental, and social criteria. This combination ensures optimal site selection, resource efficiency, and sustainable outcomes, emphasizing its importance in driving the transition to a cleaner energy future.

References

1. (2023) Visor INFOMAR - MITECO, CEDEX. http://www.infomar.miteco.es/visor.html, accessed on 01 Oct 2024
2. Adedeji PA, Akinlabi S, Madushele N, Olatunji OO (2021) Spatial location of renewable energy plants: How good is good enough? In: Vijayan S, Subramanian N, Sankaranarayanasamy K (eds) Trends in manufacturing and engineering management. Springer Singapore, Singapore, pp 1055–1064
3. Ali SCTKF (2023) Assessment of small hydropower in songkhla lake basin, thailand using gis-mcdm. Sustain Water Resour 9. https://doi.org/10.1007/s40899-022-00788-w
4. Escudero AR, Gil IG, García AM, del Socorro García Cascales M (2023) A multi-criteria decision-making tool to identify optimal locations for pv-geothermal systems for heating and cooling in industrial areas. In: EGU general assembly 2023, https://doi.org/10.5194/egusphere-egu23-1566, https://meetingorganizer.copernicus.org/EGU23/EGU23-1566.html
5. Gil García IC, et al (2020) Integración del recurso eólico marino en los sectores del transporte y climatización: estudio de transición energética en la costa este de eeuu. Ph.D. Thesis, Universidad Politécnica de Cartagena, available online: https://repositorio.upct.es/handle/10317/9164
6. Gil-García IC, García-Cascales MS, Fernández-Guillamón A, Molina-García A (2019) Categorization and analysis of relevant factors for optimal locations in onshore and offshore wind power plants: a taxonomic review. J Mar Sci Eng 7(11). https://doi.org/10.3390/jmse7110391, https://www.mdpi.com/2077-1312/7/11/391
7. Gil-García IC, Ramos-Escudero A, García-Cascales M, Dagher H, Molina-García A (2022) Fuzzy gis-based mcdm solution for the optimal offshore wind site selection: the gulf of maine case. Renew Energy 183:130–147. https://doi.org/10.1016/j.renene.2021.10.058
8. Gil-García IC, Ramos-Escudero A, Molina-García Ángel, Fernández-Guillamón A (2023) Gis-based mcdm dual optimization approach for territorial-scale offshore wind power plants. J Clean Prod 428:139484. https://doi.org/10.1016/j.jclepro.2023.139484, https://www.sciencedirect.com/science/article/pii/S0959652623036429
9. Lozano JMS (2013) Búsqueda y evaluación de emplazamientos óptimos para albergar instalaciones de energías renovables en la costa de la región de murcia: combinación de sistemas de información geográfica (sig) y soft computing. Tesis doctoral, Universidad Politécnica de Cartagena, https://dialnet.unirioja.es/servlet/tesis?codigo=51908&info=resumen&idioma=SPA, https://dialnet.unirioja.es/servlet/tesis?codigo=51908
10. García Cascales MS (2009) Métodos para la comparación de alternativas mediante un sistema de ayuda a la decisión: S.a.d. y "soft computing". Ph.D. Thesis, Universidad Politécnica de Cartagena, available online: https://repositorio.upct.es/handle/10317/9164

11. Mokarram M, Akbarian Ronizi SR, Negahban S (2024) Optimizing biomass energy production in the southern region of iran: a deterministic mcdm and machine learning approach in gis. Energy Policy 195:114350. https://doi.org/10.1016/j.enpol.2024.114350, https://www.sciencedirect.com/science/article/pii/S0301421524003707
12. Mozaffari M, Bemani A, Erfani M, Yarami N, Siyahati G (2023) Integration of lcsa and gis-based mcdm for sustainable landfill site selection: a case study. Environ Monit Assess 195:510. https://doi.org/10.1007/s10661-023-11112-0
13. Munier N, Hontoria E (2021) Shortcomings of the AHP method. Springer, Cham, pp 41–90. https://doi.org/10.1007/978-3-030-60392-2_5,
14. Pathan AI, Agnihotri PG, Patel D (2022) Integrated approach of ahp and topsis (mcdm) techniques with gis for dam site suitability mapping: a case study of navsari city, gujarat, india. Environ Earth Sci 81:443. https://doi.org/10.1007/s12665-022-10568-6
15. Puppala H, Arora MK, Garlapati N, Bheemaraju A (2023) Gis-mcdm based framework to evaluate site suitability and co2 mitigation potential of earth-air-heat exchanger: a case study. Renew Energy 216:119072. https://doi.org/10.1016/j.renene.2023.119072
16. Rahimi S, Hafezalkotob A, Monavari SM, Hafezalkotob A, Rahimi R (2020) Sustainable landfill site selection for municipal solid waste based on a hybrid decision-making approach: fuzzy group bwm-multimoora-gis. J Clean Prod 248:119186. https://doi.org/10.1016/j.jclepro.2019.119186, https://www.sciencedirect.com/science/article/pii/S0959652619340569
17. Ramos-Escudero A, Magraner T, Gil-García IC (2024) Optimized spatial tool for the implementation of ground source heat pump coupled with photovoltaic panels heating systems in urban areas. Energy Build 323:114752. https://doi.org/10.1016/j.enbuild.2024.114752, https://www.sciencedirect.com/science/article/pii/S0378778824008685
18. Ramos-Escudero G et al (2020) Gis-based analysis for geothermal energy development and potential mapping. Appl Energy 268:114–127
19. Rekik S, El Alimi S (2024) A gis based mcdm modelling approach for evaluating large-scale solar pv installation in tunisia. Energy Rep 11:580–596. https://doi.org/10.1016/j.egyr.2023.12.018, https://www.sciencedirect.com/science/article/pii/S2352484723016104
20. Saraswat S, Digalwar AK, Yadav S, Kumar G (2021) Mcdm and gis based modelling technique for assessment of solar and wind farm locations in india. Renew Energy 169:865–884. https://doi.org/10.1016/j.renene.2021.01.056, https://www.sciencedirect.com/science/article/pii/S096014812100063X
21. Shao M, Han Z, Sun J, Xiao C, Zhang S, Zhao Y (2020) A review of multi-criteria decision making applications for renewable energy site selection. Renew Energy 157:377–403. https://doi.org/10.1016/j.renene.2020.04.137, https://www.sciencedirect.com/science/article/pii/S0960148120306753
22. Spyridonidou S, Vagiona DG (2020) Spatial energy planning of offshore wind farms in greece using gis and a hybrid mcdm methodological approach. Euro-Mediterr J Environ Integr 5:24. https://doi.org/10.1007/s41207-020-00161-3
23. Sánchez-Lozano J, García-Cascales M, Lamata M (2016) Gis-based onshore wind farm site selection using fuzzy multi-criteria decision making methods. Evaluating the case of southeastern spain. Appl Energy 171:86–102. https://doi.org/10.1016/j.apenergy.2016.03.030, https://www.sciencedirect.com/science/article/pii/S0306261916303543
24. Sánchez-Lozano JM, Teruel-Solano J, Soto-Elvira PL, Socorro García-Cascales M (2013) Geographical information systems (gis) and multi-criteria decision making (mcdm) methods for the evaluation of solar farms locations: case study in south-eastern spain. Renew Sustain Energy Rev 24:544–556. https://doi.org/10.1016/j.rser.2013.03.019
25. Yildiz S (2024) Spatial multi-criteria decision making approach for wind farm site selection: a case study in balıkesir, turkey. Renew Sustain Energy Rev 192:114158. https://doi.org/10.1016/j.rser.2023.114158, https://www.sciencedirect.com/science/article/pii/S136403212301016X

MIX
Papier aus verantwortungsvollen Quellen
Paper from responsible sources
FSC® C105338

If you have any concerns about our products,
you can contact us on
ProductSafety@springernature.com

In case Publisher is established outside the EU,
the EU authorized representative is:
**Springer Nature Customer Service Center GmbH
Europaplatz 3, 69115 Heidelberg, Germany**

Printed by Libri Plureos GmbH
in Hamburg, Germany